BATTLEFIELD EVOLUTION
WORLD AT WAR

Credits

Author
Agis Neugebauer

Developer
Matthew Sprange

Editor
Nick Robinson

Cover Art
Ben Wootton

Logo
Giles Meakin

Proofreading
Ted Chang, Bill Swansong

Interior Illustrations
Giles Meakin

Playtesting
Michael Hellmig, Tom Weiss, Dominik Cenia, Jason Coffey, Dan Tuziak, Steve Holt, Ted Pawlowski, Rick San Agustin, Tom Adams, Marcin Budzinski, Krzysztof Kowalczyk, Gustav Barczak, Jaroslaw Grabowski, Alexander Fennell, Richard Zanetti, Dennis Perlot, Chris Jaques, Robert Seitz

Miniatures and Terrain
Bolt Action Miniatures, Tablescape, Artizan Designs, Dragon, Italieri, Sands (S&S Models), TQD Castings, Trumpeteer, West Wind Productions

Miniature Painters
Agis Neugebauer, Tom Weiss, Big Gazza, Darren Linnington and Simon Bargery

Copyright Information

Printed in the USA.

CONTENTS

THE LATE WAR, EUROPE

World War II enveloped almost the entire globe, and affected the lives of billions. Perhaps more than sixty million people were killed as a direct result of the war, and maybe as many as half of these were Russian. The exact numbers of dead, soldier and civilian, will never be known.

As 1944 turned, the war was in its last throes but the fighting was reaching its highest pitch. The Allies, having suffered greatly in the first years of the war, could sense the war was coming to a close. Preparations in the west were being made for the largest beach landings in history, and the liberation of France would be a prelude to the liberation of the rest of Western Europe. In the east, Russia had suffered more than any other nation, but its great strength was finally being marshalled to roll back the Nazi war machine to its own borders.

Monte Cassino

D-Day

Eastern Front

Falaise Gap

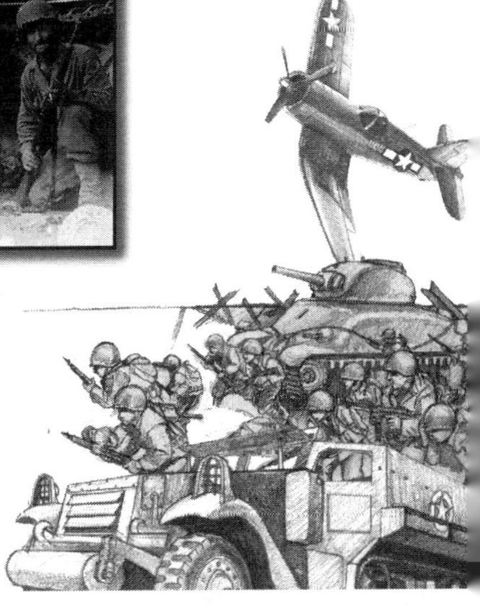

The Late War, Europe

In Germany, there was a very real fear of defeat, carried by the knowledge that this time there would be no civilised agreements ending hostilities. Barbarians were coming from the east, and the west wanted vengeance for years of terror and occupation. The invaders would march through Germany's cities and culture. The German people were preparing to fight for their identity and survival, by increasingly more desperate means.

In the end, there would be a New World Order, but it was not to be the one Adolf Hitler had imagined. Europe would be liberated, new nations would rise to the world stage, and the causes of both democracy and communism would be seen to be the victors over fascism, although at the cost of a new, colder war in the following decades.

However, at the start of 1944, there was still along road to travel before Final Victory would be realised.

Battle of the Bulge

Operation Market Garden

Battle of Berlin

VE Day

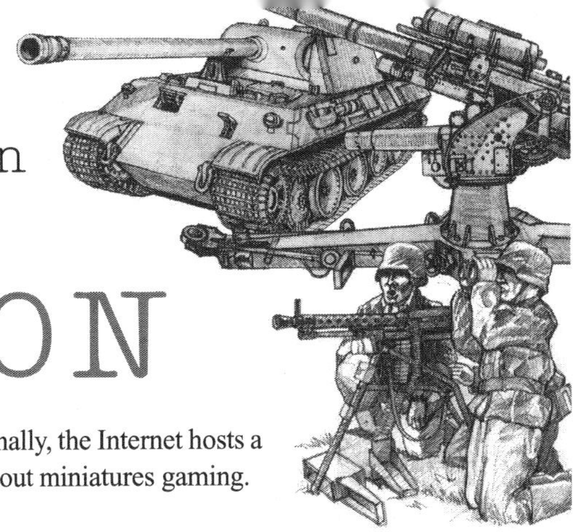

INTRODUCTION

Welcome to the war, Soldier!
Battlefield Evolution is a fast-paced and easy to learn miniatures game, with enough tactical depth to keep you hooked for years. You will command platoon and company-sized forces, and attempt to overwhelm your opponents with tactical skill and a dose of luck!

Miniatures Gaming

Miniatures gaming is a hobby in which players collect armies of model soldiers and vehicles, and then take part in games against like-minded opponents. Played on a table featuring terrain ranging from open fields to burnt out city streets, miniatures gaming allows you to act as a general, directing your forces in a cunning battle plan designed to bring you total victory!

You can think of it as chess but;

- With freeform movement – you are not restricted to squares.
- With random elements – troops pinned down by enemy fire are likely to be in real trouble, but they might surprise you and start fighting back.
- With far prettier models!

Using the rules of a miniatures game – aided by a few models, some dice and a tape measure – you will be able to lead your armies across the battlefield, dodging a hail of incoming fire, before unleashing your finest troops against your enemy.

The joy of miniatures games like this one is that they are flexible (no two games will play out quite the same way), challenging (with all the different models you can use there a lot of different tactics possible) and visually appealing (nothing beats the clash of two forces over some nice model terrain). Miniatures gaming is a fascinating hobby that appeals to all ages and walks of life. If these are your first steps into the hobby then you are doubly welcome – you have just found something that may well keep your interest for a long time.

Your local hobby store can be a great resource for learning about the game. Many stores run their own clubs, tournaments and ongoing campaigns, which offer a great chance to find opponents. Additionally, the Internet hosts a cornucopia of sites and forums about miniatures gaming.

Battlefield Evolution: World at War

The Second World War was the most widespread conflict in human history, spanning almost every continent and affecting the lives of billions. The fortunes of nations rose and fell, and battlefield technology accelerated at a heavy pace.

Battlefield Evolution: World at War concentrates on the late period of the war (1944 onwards) and the European theatre in an effort to do justice to the forces involved. Within these pages, you will find the core rules of the game, as well as British, American, German and Soviet army lists covering the main forces of each nation. Future supplements and expansions will explore other theatres and specific campaigns, as well as different types of warfare and new units.

Needed For Play

In addition to this rulebook, you will also need several other items to begin playing this game. While additional materials such as terrain pieces and counters for various conditions are also useful, they are optional and not required for basic games.

- Two or more players.
- A playing surface, such as a tabletop or unobstructed floor space.
- A selection of miniatures.
- A tape measure or ruler of a minimum 12 inches in length.
- Several six and ten-sided dice (at least two of each).

Rolling Dice

This game uses two kinds of dice – the ordinary, humble six-sided dice and the chunky, exotic ten-sided dice. You will find it beneficial to have perhaps a dozen D6s on

hand as it greatly speeds up play. D10s are used more rarely and you should not need more than a couple of these. The size or variety does not matter – they will suffice as long as they are six-sided (numbered 1 through 6) and ten-sided (numbered 1 through 10).

You will sometimes be instructed to read the results of dice rolls in different ways. If you are simply told to roll a dice in the rules, roll it as normal and read the number or pips on the top. However, you will sometimes see references telling you to roll 4xD6, 6xD6, 2xD10+1, etc. These are dice shorthand references.

A number before a 'xD' tells you how many dice you need to roll. The number after a 'D' tells you whether to roll D6s or D10s. For example, 4xD6 means 'roll four six-sided dice' (you can see why we shorten it to 4xD6). If there is a '+' after the dice add the number to each dice rolled – so 4xD6+1 would tell you to roll four six-sided dice and add one to the resulting score on each.

Some rules may ask you to add dice rolls together, such as 2xD10+D6. In this case, you would roll two D10s, one at a time, adding a separate D6 roll to each (the D6 is rolled once, with the result being added to both the D10 rolls).

Where multiple dice of the same type are rolled and added together it will be noted as the number of dice only, without the 'x' symbol. For example, 2D6 (giving you a total of 2 to 12), 3D6 (3 to 18) or 2D10 (2 to 20). A numerical addition (such as the 2 in '3D10+2') is added to the total value. Thus, if you were intended to roll four six-sided dice, add their total together and then add a fixed value of 1, it would be written as 4D6+1.

Natural 1s

These rules assume that an unmodified roll of '1' on a D6 or D10 is automatically a failure. Even if subsequent bonuses raise the result to above 1, it is still a failed roll. This is referred to as rolling a 'natural 1'. The natural 1 rule applies to all rolls in this game unless specifically mentioned otherwise.

Re-Rolls

Some special situations may call for you to re-roll a dice. This simply means you ignore the first result rolled and roll the dice again. All modifiers or conditions that affected the first roll are also present and must be accounted for on the re-roll. You must always accept the result of the second roll, even if it was worse than the first – re-rolls can be used to get you out of a tricky situation but they are never guaranteed! Also, you may only ever re-roll a specific dice result *once*, even if you have multiple opportunities to re-roll a dice result.

Turns

Players take turns to play Battlefield Evolution. Throughout these rules, a turn is used to describe the process a player goes through before his opponent takes his turn and repeats that process. A Game Turn is two of these 'player' turns.

Pre-Measuring Distances

You may always 'pre-measure' distances in Battlefield Evolution, allowing you to check to see if an enemy is within range before making the decision to fire, for example.

Preparing for Play

Once you have your army and an opponent, there are just a couple of things you need to do before battle can commence. You will need a flat playing surface, such as a kitchen or dining room table – though many players invest in a dedicated gaming table, complete with modelled terrain.

Your playing surface should have plenty of terrain placed upon it. You can use piles of books or boxes to simulate buildings, walls and hills, though many players soon progress onto model houses, woods, factories, ruins and so on.

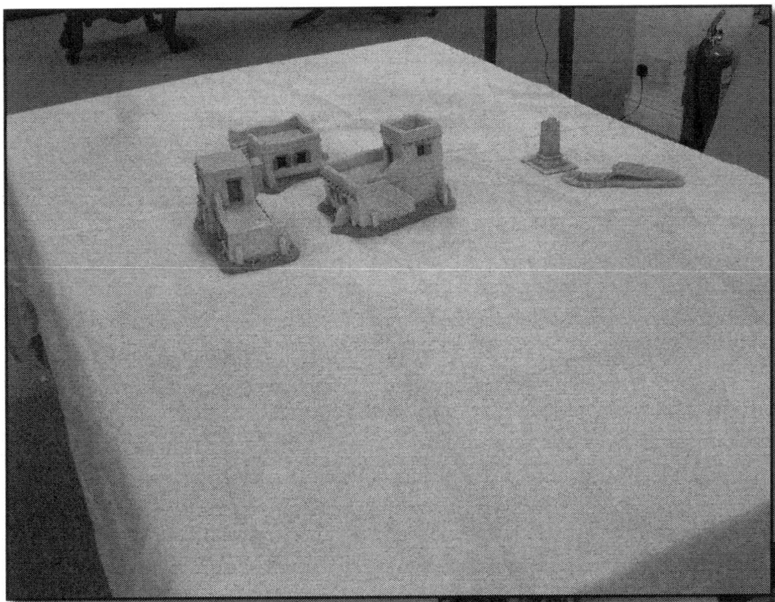

A basic tabletop terrain can made by covering a table with a coloured sheet and using commercially available terrain

A more adventurous table, made from polystyrene with a selection of modified terrain pieces

This advanced battlefield has been almost completely scratch built, showing what can be achieved with even a modicum of modelling skill

BUILDING AN ARMY

Before you can start fighting huge and involved battles, you will need to build an army. An army can be as small as a single squad or feature over a hundred soldiers, supported by vehicles, command units and aircraft. You can go on to field a whole company of troops with additional support, limited only by the size of your wargaming table!

Every unit in the game is assigned a points value, as shown in the army lists on page 41 onwards. As you collect units, add their points values together, as this will show you just how large and powerful your army is.

When you come to play a game, the force you use may only be a fraction of your collected army. This is because the points system is used to ensure a fair match between yourself and your opponent. Two forces of the same points value should be roughly equal in power, meaning it will be your skill and tactical ability that will be the deciding factor for victory.

We recommend your first battles are played with forces of no more than 2,000 points each. This will give you a game that will last up to two hours.

Ultimately, there is no limit to the size of armies that can be played, and you will find that even games involving thousands of points can still be completed within a fairly short period of time, so long as both players are familiar with the rules. The only real restraint is the size of your gaming table, and your ability to actually place all your models on it. However, even games involving armies of 6,000 or 8,000 points can still be realistically played on an average 6 ft. by 4 ft. gaming table, though larger forces may require you to use a wider table or even the floor!

A powerful American army advancing through the snow

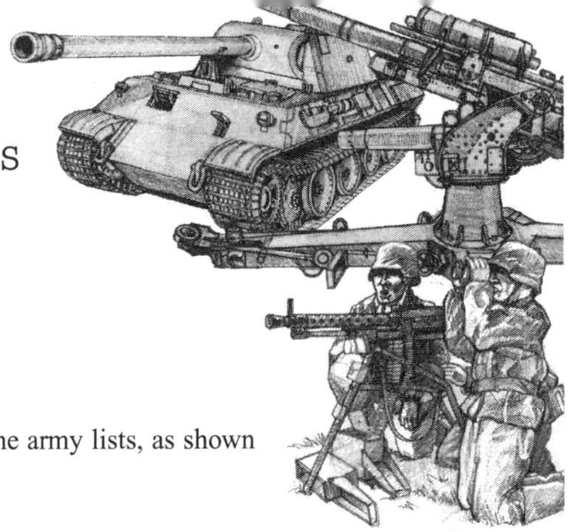

UNITS

Models are grouped into units, which act together on the battlefield. A unit may be several models, such as a squad of soldiers or a single larger or important model, such as an armoured vehicle.

A unit is defined by its entry in the army lists, as shown below.

Grenadier Squad – 160 points

Well trained to attack using speed, German Grenadiers were often found where the fighting was at its fiercest.

Type	Size	Move	Close Combat	Target	Save	Kill	Traits
Unteroffizier	1	4”	D6+1	4+	6+	6+	-
Obergefreiter	1	4”	D6+1	4+	6+	6+	-
Soldier	1	4”	D6+1	4+	6+	6+	-

Team 1: Unteroffizier (unit leader) with a MP-40 SMG, 3 Soldiers with Kar-98k rifles, Soldier with an MG-42 LMG.

Team 2: Obergefreiter with a MP-40 SMG, 4 Soldiers with Kar-98k rifles.

Teams: The Obergefreiter may act as unit leader to Team 2, splitting off to act as a separate unit.

Tank Assault: Infantry in this period used a variety of grenades or mines to assault armoured vehicles. All members of the unit are assumed to carry Anti-Tank Explosives. These may be placed on any model of Size 2 or greater with a Charge action if the target ends up within the Lethal Range of the charging unit. They will explode at the end of the charging unit’s next turn.

Options: Up to 1 Soldier of Team 2 may be equipped with an MG-42 LMG for +40 points. Up to 1 Soldier of either Team 1 or 2 may be equipped with a Panzerfaust in addition to his rifle for +30 points. Drop Team 2 for – 60 points. A Grenadier Squad may have an Opel Blitz Medium Truck for +50 points, an SdKfz 251/1 halftrack for +80 points, or up to 2 SdKfz 250/1 Neu halftracks for +70 points each.

Title: The name of the unit.
Points: The total points value of the unit. These are added together to obtain the points value of your whole army.
Type: The name of individual models in a unit.
Value: The individual points value of each model in the unit.
Size: The Size score of each model in the unit. This is used for a variety of purposes, such as measuring how many models can fit inside a transport.
Move: How far each model in the unit can normally move across the table in inches with a single Move action. Movement is explained on page 9.
Close Combat: How many dice each model in the unit rolls when it charges an enemy model in close combat. Close combat is explained on page 22.
Target: The number an enemy has to roll equal to or greater than on a single dice when making an attack against a model in this unit.
Save: The number that must be rolled if a model in the unit is hit by an enemy attack. If a second score is present

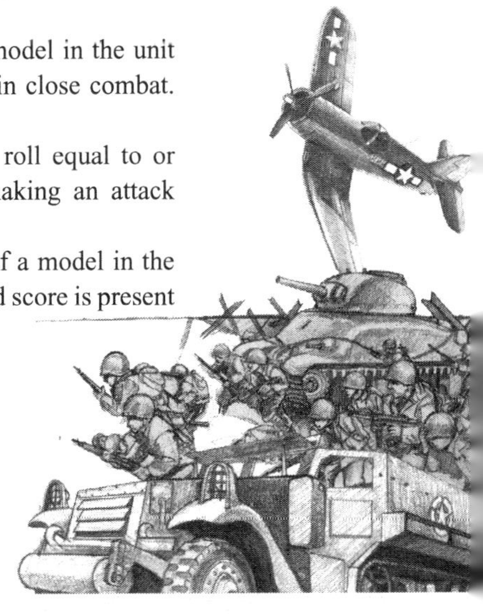

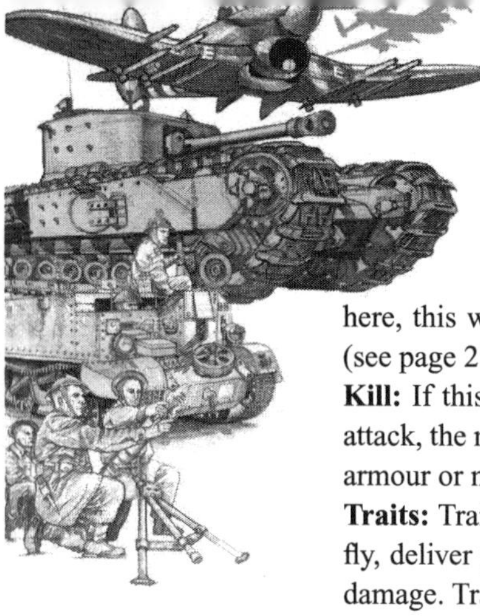

here, this will mean the model can make a Dodge save (see page 21 for details).
Kill: If this number is rolled when an enemy makes an attack, the model is immediately removed, whether it has armour or not – the attack was just too powerful!
Traits: Traits allow models to do special things, such as fly, deliver powerful attacks or withstand a great deal of damage. Traits are explained on page 27.
Unit or Team: Which models are included in the unit or team as standard, before any options are taken, along with the weapons they carry. A unit that is split into multiple teams this way only counts as a single unit choice. If the piat team from a British Infantry Command Squad (see page 41) was split off as a seperate team it would not count as an additional support unit, for example.
Special Rules: Many units have special rules not covered by Traits that allow them to do things in the game that other units cannot even attempt. Special rules also allow you to do things not normally permitted in the rules!
Options: Any options the unit can take are listed here, along with any relevant changes in its Value.

Unit Leaders

Every unit in the game has a unit leader, as listed in its army list entry. If a unit consists of only one model, then that model is assumed to be the unit leader.

All models in the unit must end every action within 6" of their unit leader. This is the unit leader's Command Range.

If a model is ever more than 6" away from its unit leader, then it will be out of command. The entire unit will be out of command if it loses its unit leader and is reduced to less than half the models it started the battle with.

If a unit loses its unit leader but has not been reduced to less than half the models it started the battle with, designate any model to become a temporary unit leader. This model is not as experienced as the original unit leader, but knows enough to get the unit moving. All models must remain within 6" of the temporary unit leader – any who do not will be out of command. If the temporary unit leader is killed, you may immediately nominate another model to take over, so long as the unit is still at half strength or greater.

The entire unit will still go out of command if it is reduced to less than half strength, even if a temporary unit leader is present.

Models that are out of command may only make reactions (see page 25).

A unit leader of one unit may not act as the unit leader of another, though some models may have special rules that do allow them to take command of other units.

If a model becomes separated from its unit by any means and is not within 6" of its unit leader, then it may only make reactions as stated above. As soon as the model's unit leader completes an action within 6", the model may rejoin the unit and be within command once more.

Facing

Many units on the battlefield are fast-moving and agile. In most cases, you do not need to worry about which way a model is pointing when you move it, as it is assumed it can turn quickly enough to face any threat and target any enemy.

Some models (such as armoured vehicles) have special rules that require you to know where it is facing. Some weapons, for example, may only be permitted to fire into one facing, or the unit may have weaker armour to the sides and rear that enemies can take advantage of.

The diagram below shows a unit and its Front, Side and Rear facings.

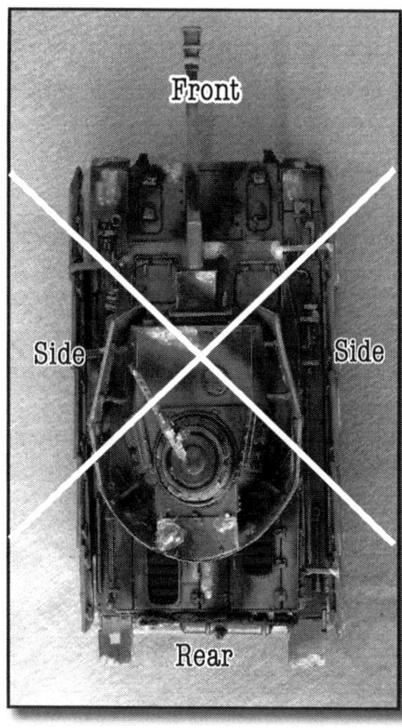

A US Bazooka team goes Panther hunting

US Infantry dismount from their M3 Halftrack

British Command, HMG and Sniper take position in and around a bunker

A British Six Pounder Anti-tank Gun

Units

British Infantry Section

German tanks with supporting infantry advance

A Soviet Desant Squad

THE GAME TURN

A battlefield is a confusing place, with many things happening at once. In order to make sense of things, games take place over several turns. The player who set up his army first will take the first turn. In your turn, you will have a chance to move and fight with each unit you have on the table. Once you have finished, your opponent then has his turn to move and fight with his units. This continues until the battle is complete and a victor is declared!

Actions

Whenever you decide to do something with a unit in your army it will have two actions and every model within the same unit will perform the same action. There are four types of action.

Move: Allows the unit to walk, run, drive or otherwise travel across the battlefield.
Shoot: Allows the unit to begin gunning down the enemy!
Charge: Allows the unit to rush into close combat to tear the enemy apart!
Ready: Allows the unit to prepare itself to do something unusual, like load a complicated weapon or plant a bomb.

A unit can do any combination of two actions. For example, a squad about to be overrun by the enemy might choose two Shoot actions. Their allies trying to flank the enemy might choose two Move actions, while another squad near the frontline might choose a Move action followed by a Shoot action.

A unit must complete both of its actions before you can start making actions with another unit. A unit need not take one or both actions if you do not want it to – you are never forced to take an action.

Over the next few pages, we take a look at these different actions more closely. The important thing to remember is that each unit gets two actions of your choice in a turn. Once all your units have taken two actions, your turn is over and your opponent does the same thing.

Victory and Defeat

The game will continue until either one side gains victory by achieving its objectives or when a certain number of turns have been completed.

MOVE ACTIONS

Every unit in Battlefield Evolution has a Move score (found in its army list entry) which shows how far its models can travel across the table, in inches, with a single Move action. A model need not move in a straight line and can make any number of turns to face in any direction you wish.

A Move action does not have to be taken to its full possible length and a unit can travel any decided distance up to its maximum value. Models may move through any friendly units freely but may not move through an enemy model unless they have a greater Size score.

All models in a unit must end their movement within the 6" command range of their unit leader.

German infantry advance

Terrain

Terrain in Battlefield Evolution is anything that is placed on the table that is not a model – this can be anything that is part of the landscape, such as buildings, woods, rivers and hills. You will quickly find that the more terrain you use, the better the game, as your units will have a chance to leap behind walls and dodge behind buildings in an effort to avoid incoming fire. If you just play on a flat 'desert' you will soon discover your troops get mowed down by the first weapon that gets fired!

There are three different types of terrain:

Clear Terrain: Includes roads, low hills and shallow streams – anything that appears to be easy to move over. Clear terrain will not hinder movement at all, and so no reduction in the Move score occurs.

Difficult Terrain: Includes steep hills, woods, and marshes – anything that looks like it would be difficult for most units to cross. Units moving through difficult terrain will find their progress hindered, counting every 1" moved as 2" (effectively halving your movement while on the terrain).

Impassable Terrain: Includes tall buildings, cliffs and quicksand. This will block all normal movement, and no model without special rules that ignore terrain may cross it.

Size

Every piece of terrain has a Size score, just as models do. This will be equal to the terrain's height (or depth in the case of crevasses or liquid sources) in inches, rounding to the nearest whole number.

Models of greater Size than the terrain may completely ignore it, counting it as clear.

Cover

Some terrain will provide cover for models that take advantage of it. This is discussed more thoroughly under Shoot actions on page 19.

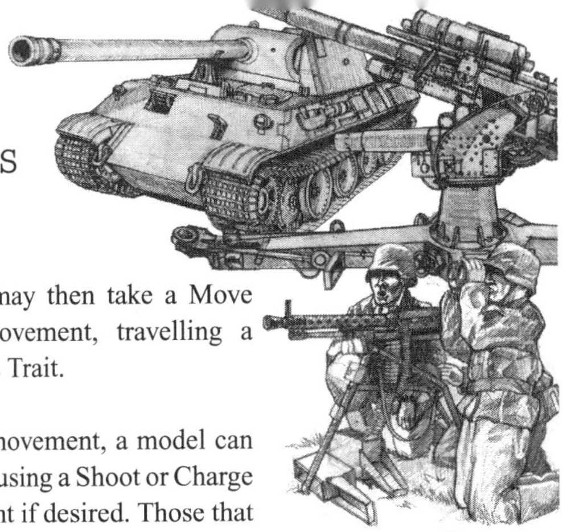

Structures

While structures (the term used for buildings) are a type of terrain, they use their own rules as it is possible to move inside a building, travel between floors and turn them into temporary strongpoints. They are covered later on page 34.

Special Movement Modes

Some models are able to use special movement rules to drive, fly or use some other special means of getting around the battlefield. Each special type of movement is a kind of Trait and will be listed in the unit's army list entry if it has it.

These special types of movement may only be used by units if they take a Ready action first, which represents the model taking to the air, preparing to climb, revving engines and so forth. The unit may then take a Move action, and use the special movement, travelling a distance up to that specified by its Trait.

When making a special type of movement, a model can also execute an attack as if it were using a Shoot or Charge action at any point in its movement if desired. Those that are flying or gliding are assumed to be a number of inches above the table, or whatever terrain they are flying over, equal to their Size.

Note that while a model *can* use its special movement mode to get around, it does not have to – for example, tracked and wheeled models have a lower 'safe speed' to grind along at more or less walking pace and an agile climber can walk along on the ground. This is represented by using a normal Move action.

SHOOT ACTIONS

Simple manoeuvring will not win you a battle – you have to destroy the enemy as well, and automatic weaponry is always a good way of doing that!

Weapons

The descriptions of every weapon an army uses can be found in the appropriate army list. An example is given here;

Weapon	Range	Damage	Traits
MG-42 LMG	48"	2xD6+1	Auto, Prone

Weapon: The name of the weapon.
Range: How far the weapon can shoot.
Damage: How many dice are rolled against a target.
Traits: This lists any special Traits the weapon possesses. See page 27 for more details.

Special Ranged Weapons

Some weapons may not have a Range score. Instead, one or more delivery modes may be stated. The delivery modes are Dropped, Launched and Placed.

Dropped munitions are released by Air Units. See page 29 for more details.

Launched munitions can be fired by a launcher system (the weapon's description will specify which launch system must be used). A range in inches will also be given. If a model does not have the requisite launch system, it may not launch this munition, though it may use a different delivery mode.

Placed munitions can be planted using the rules found in their army lists.

Fire Zones and Line of Sight

Before you can shoot at an enemy, the models in your own unit must have a Line of Sight to their intended targets.

In all cases, Line of Sight is traced as a straight line from the centre of each of the firing models to any part of a potential target model. Firing models must also be able to trace a Line of Sight to the centre of a Fire Zone, as detailed below.

A Fire Zone is simply a 3" radius circle. The centre point of a Fire Zone must be an enemy model.

When you roll dice for your unit's shooting, you can only attack models completely or partially within this Fire Zone that are also within Line of Sight. All models in the unit must use the same Fire Zone – any models that cannot draw Line of Sight to the centre of the Fire Zone may not join in on the attack, whether or not they can draw Line of Sight to any individual models within the Fire Zone.

Line of sight has three possible categories: Clear, Obscured and Blocked.

Note that attacking models within the same unit will never block or obscure Line of Sight for each other

A 3" Fire Zone

– effectively, you can shoot 'through' models in the same unit, as it is assumed they will manoeuvre out of each other's way. In addition, models within a Fire Zone will not block Line of Sight to other models within the same unit. This means one man in a squad cannot block or obscure Line of Sight to the other men in the squad – they are all potential targets when under fire!

Models not in the same unit *can* block or obscure Line of Sight.

Every model with a weapon can use it to attack any enemy that it can draw Line of Sight to and is within range. Range is always measured to the centre of the Fire

Clear

Clear: No terrain (or terrain that does not block Line of Sight, such as a river or road) exists between attacker and target. No effect, shooting is possible.

Obscured

Obscured: Terrain interrupts Line of Sight but both models can still partially see one another (that is, terrain hides any part of the target model). The target gains a +1 bonus to its Kill score.

Blocked

Blocked: Terrain interrupts Line of Sight so neither model may see the other. No Line of Sight possible.

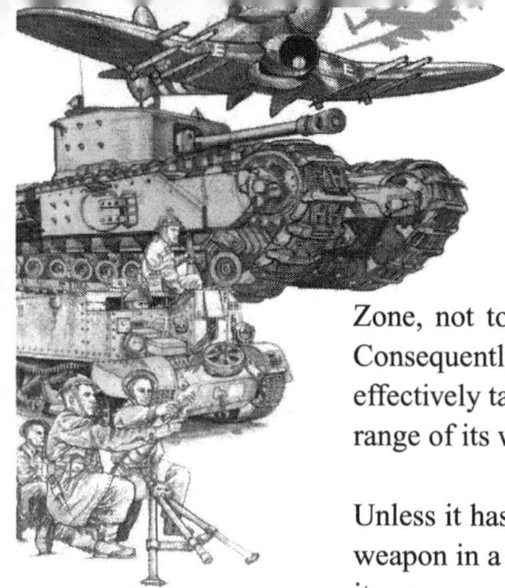

Shoot Actions

Zone, not to each enemy model within the Fire Zone. Consequently, due to a Fire Zone's 3” radius a model can effectively target an enemy 3” further than the maximum range of its weapon.

Unless it has special rules, each model can only use one weapon in a Shoot action, no matter how many weapons it possesses.

Every weapon has a set number of Damage Dice, which are rolled to see if a target has been hit and wounded or killed.

Elevated Models

It is quite common for some models within a Fire Zone to be on a different level of a building to others, or even just standing on some rocks or ruins! So long as a model is within 3” (6” for weapons with the Auto trait) of the level the model the Fire Zone is placed upon, it can be included within the Fire Zone.

The Bazooka-armed trooper is within the Fire Zone despite being an elevated model

Shooting

Roll Damage Dice for your entire unit. Discard all Damage Dice that roll a natural ‘1’. These represent missed shots or ammunition jams.

The highest scoring Damage Dice (including any bonuses) must be allocated to the model closest to the shooting unit within the Fire Zone, the next highest scoring Damage Dice to the next nearest model, and so on. If all models within the Fire Zone have been allocated a Damage Dice, then you are free to allocate any remaining Damage Dice to models as you see fit, placing multiple dice on specific models that you deem especially dangerous, if you so wish.

Note that friendly models within a Fire Zone are potential targets as well! However, a model may never be caught

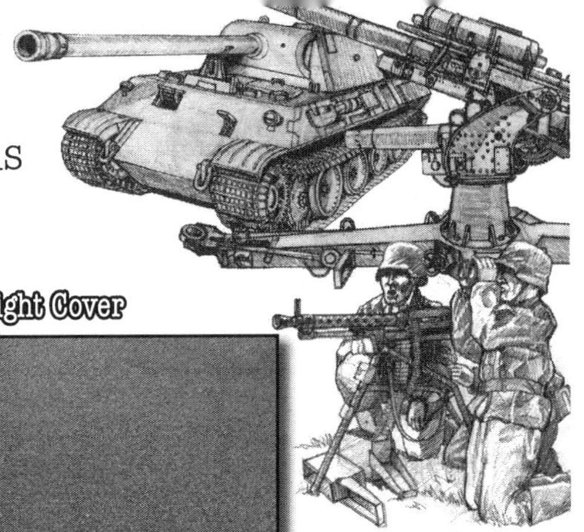

within a Fire Zone that it helped create itself (in other words, models cannot shoot themselves or members of their own unit!).

Also be aware that it is perfectly permissible for a Fire Zone to overlap more than one unit.

All models have two scores which you need to aim for when attacking them – a Target score and a Kill score. If you roll equal to or greater than the model's Kill score, it is automatically removed from the battlefield. You have dealt it a grievous injury or even killed it outright!

If you merely roll equal to or more than a model's Target score, then it will have a chance to use its Armour to protect itself. Your opponent must roll one six-sided dice and score equal to or greater than his model's Armour score in order to survive. A roll of 1 for Armour is always a failure.

Cover

Most types of terrain will also provide Cover for models, effectively shielding them from shooting attacks. Different terrain will provide different amounts of cover – a thick reinforced wall, for example, will provide better protection than tall grass!

There are three different types of terrain that can provide Cover, as listed below.

Type of Cover	Example Terrain	Bonus Provided
Light	Tall grass, crops, marshland	+1 Kill score
Medium	Woods, ruins, walls	+1 Armour Save, +1 Kill score
Heavy	Fortified walls, rocks	+2 Armour Save, +2 Kill scorc

A model has to be touching (on the opposite side of the terrain to the firing model) or actually completely within suitable terrain to take advantage of Cover. Otherwise, the terrain can only obscure or block Line of Sight.

For the purposes of Line of Sight only, a firing model may 'see through' any Cover within a number of inches equal to its Size score. It may also 'see through' any Cover within a number of inches to the target model that is equal to the target model's Size score. Models, if within terrain, may always claim Cover rather than just have an obscured Line of Sight.

A fence provides Light Cover

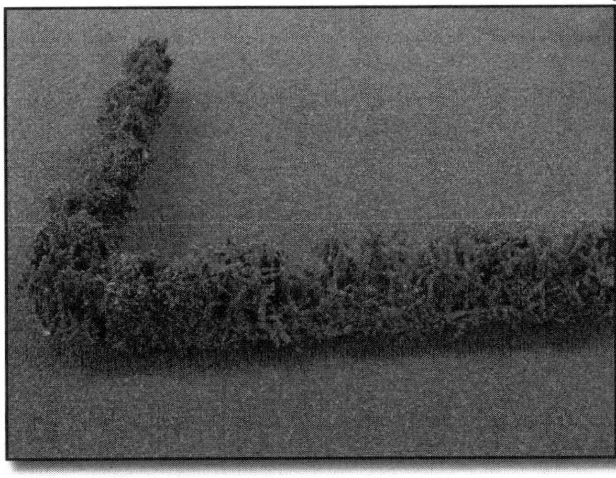

A wall provides Medium Cover

These rocks provide Hard Cover

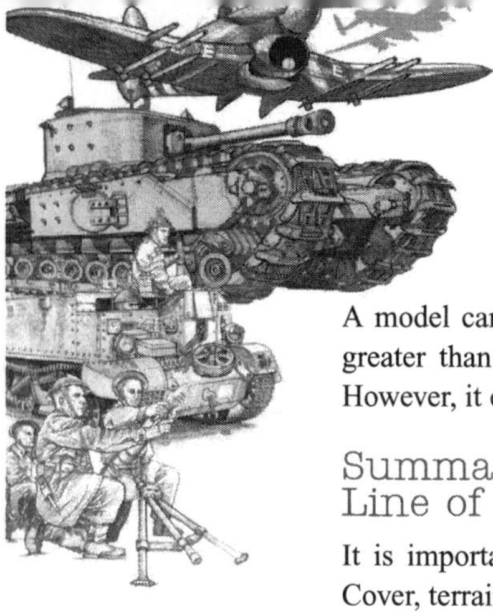

A model cannot benefit from Cover if its Size score is greater than that of the terrain it is within or behind. However, it can still count as being Obscured.

Summary of Cover, Terrain and Line of Sight

It is important to understand the relationship between Cover, terrain and Line of Sight.

- Terrain can block or obscure Line of Sight.
- In some circumstances, terrain can instead provide Cover.
- Being able to trace a clear Line of Sight does not necessarily mean a model is deprived of Cover.
- A model cannot be Obscured and gain Cover at the same time.

Note that if some models shooting into a Fire Zone can draw a clear Line of Sight to an enemy model or deprive him of Cover, while other models on the same attacking unit cannot, you should roll your Damage Dice separately for them.

Splitting Fire Zones

A unit can opt to split fire by declaring extra Fire Zones if you wish. In effect, this is the act of spreading an attacking unit's fire wider to catch more targets or shooting in different directions. The downside to this tactic is that the unit's firepower is weakened by being spread in this way.

Any two Damage Dice of your choice must be discarded for each extra Fire Zone created and the remaining Damage Dice must be divided between the two or more Fire Zones before any of them are rolled.

Each individual weapon may only allocate dice to a single Fire Zone. You can only split the firing of a unit, not a single weapon!

The American Bazooka targets tank, while the rifles target infantry

Damage Bonuses

Some weapons have a damage bonus because they are extremely accurate, have a high rate of fire or possess exceptional penetration. If a damage bonus applies, you may need to roll different coloured dice for that weapon when firing. Each damage roll adds the damage bonus before comparing its value to the Target and Kill characteristics of the attacked model.

For example, a 3xD6+1 damage weapon adds one to the score on each of the three dice rolled for it.

Some attacks receive a random damage bonus by rolling an extra dice (this being the differently coloured dice). Remember that this is not a Damage Dice so does not count as an extra hit on its own, nor is it ignored if it rolls a 1; the bonus dice is purely there to boost the weapon's Damage Dice scores. Only one such bonus dice is ever rolled during an attack, regardless of how many Damage Dice the weapon rolls, with the result being added to each of the Damage Dice.

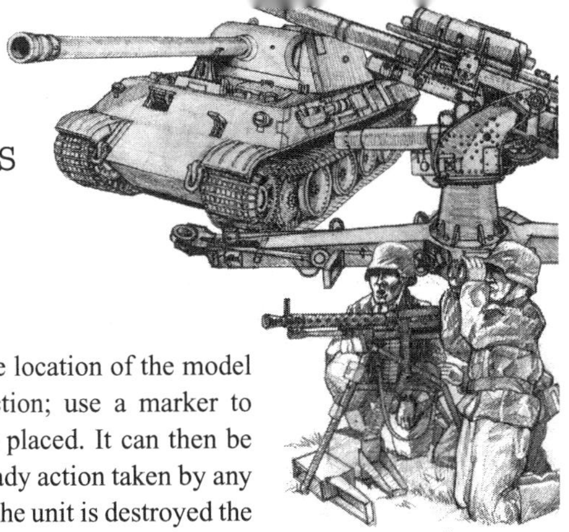

Fire Arcs

Some models, usually large and ungainly units, have limited fields of fire, and are unable to react quickly when enemies start crowding around them. Some weapons on such units will have a designated fire arc detailed in parentheses in their army list description. Such weapons may only place Fire Zones within their fire arc.

FaF: Forward
FaL: Left
FaR: Right
FaA: Aft

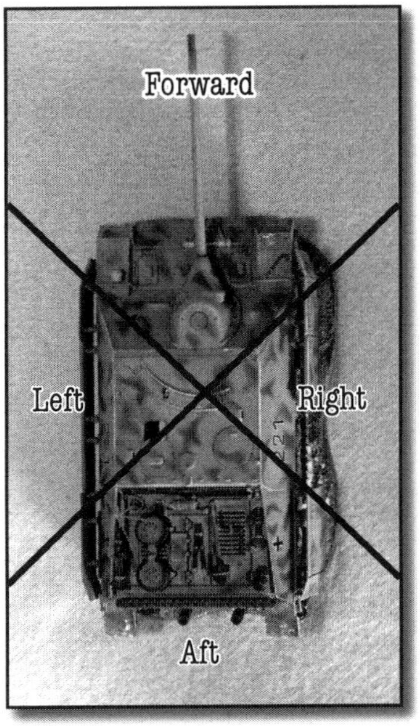

A German Jagdpanzer IV has a limited forward fire arc for its main armament.

Destruction of Large Models

If a model of Size 3 or greater is destroyed, it should be left on the battlefield as wreckage, counting as terrain and possibly providing Cover or obscuring Line of Sight. You may find it visually appealing to either turn the model upside down, cover it with cotton wool to represent smoke, or even create your own 'destroyed vehicle' terrain.

Dodge Saves

Some models will have two scores listed for their Armour – the second will be a Dodge save, allowing the model to nimbly jump out of harm's way in the last instant.

A model's Dodge save is never modified and will always be a straight roll on a D6. An Armour save, for example, can be modified by a Piercing weapon, but Dodges cannot.

Dodges can be used to try to save a model from harm even if their Kill score has been rolled, but you may only make one save against each attack, and so must choose between using Armour or trying to Dodge.

Remote Weapons

A Remote weapon is Placed at the location of the model carrying it by using a Ready action; use a marker to show the weapon's location once placed. It can then be triggered with any subsequent Ready action taken by any model in the unit that placed it. If the unit is destroyed the Remote weapon is lost as well. These two Ready actions can be separated by any number of other actions – this normally involves the placing model taking Move actions to get out of the Remote weapon's Lethal Zone!

Artillery Weapons

Artillery fire allows some weapons to attack targets that are hidden and not in Line of Sight, effectively raining fire down from the sky. Mortars and some missiles are examples of weapons that can be fired in this manner.

Note that a weapon with the Artillery trait may always be fired normally, as detailed earlier in this chapter. These rules are only used when the player wishes to take advantage of the Artillery trait.

Impact Point and Inaccuracy

Before using the Artillery trait, the unit firing the weapon must spend a Ready action in preparation of the attack. You may then select an Impact Point anywhere between half the weapon's range and its maximum, which is where the shot will land – this need not be in Line of Sight as with normal shooting attacks.

The opposing player rolls one D10 and moves the Impact Point this many inches in any direction. That attacking player then rolls another D10 and moves the Impact Point this many inches directly back towards its original position. If the impact point is moved off the table at any stage it is assumed the fire has missed wildly or the fire mission has not gone ahead.

This will likely mean the Impact Point moves a little. Artillery fire is known for being a little inaccurate!

Once the final Impact Point has been determined, roll the weapon's Damage Dice against every model either fully or partially within the weapon's Lethal Zone.

A unit under attack from Artillery fire may never benefit from Cover or be Obscured. Artillery attacks always conducted as if fired from an elevated position.

CHARGE ACTIONS

If an enemy is in cover he can be very hard to defeat – sometimes you might find it useful to simply charge and root him out with close combat! While this can be exceedingly dangerous (you will present an easy target as you come rushing in), it is usually decisive, one way or the other.

When a Charge action is made, you may move your models up to their Move score towards an enemy unit. Note that your models must still end the Charge action within the 6" command range of their unit leader, and that terrain will slow or halt their progress as with normal Move actions.

If you succeed in getting within Lethal Range of any enemy models with one of yours, you immediately fight in close combat.

Lethal Range

Every model has a Lethal Range in which it can make close combat attacks. This is an area measured from the model's centre to a distance equal to its Size score.

Fighting in Close Combat

Roll the Close Combat Dice for all the models in your unit that have at least one enemy model in their Lethal Range.

As with shooting, if you beat your enemy's Target score, you will force an Armour roll. If you beat its Kill score, you will destroy outright. Discard all Close Combat dice that roll a natural '1'.

The highest scoring Close Combat Dice (including any bonuses) will be allocated to any enemy model you choose within one of your own models' Lethal Range, the next highest scoring Close Combat Dice to another enemy model, and so on. Once all enemy models within your models' Lethal Range have been allocated a Close Combat Dice, then your opponent is free to allocate any remaining Close Combat Dice to models as they see fit, placing multiple dice on specific models, if they so wish.

German infantry Charge into contact. Once all German models have moved the German roll Close Combat dice for every model with an American model in its Lethal Range

Push Back

After close combat has been completed, move the model with the smallest Size score directly away until it is out of its opponent's Lethal Range. If both models have the same Size score, move the model that was attacked 2" directly away – other models should be moved out of the way if necessary.

This may cause a model to be pushed back off a high point, such as a cliff or tower. Measure the distance fallen, and then consult the table below. Models with the Climb, Jump or Hover Traits reduce the distance fallen by 3". You may not make a Save of any kind against falling damage.

Distance Fallen	Damage Taken
Up to 3" fall	Roll 1xD6 damage die against the model.
Up to 6" fall	Roll 1xD10 damage die against the model.
7"+ fall	Roll 1xD10 for every 6" fallen. Add the model's Size as a damage bonus.

Pushed Back Into Impassable Terrain or off the Table

Being forced into a corner or out of a field of engagement is another way to lose models. If a model is Pushed Back into impassable terrain or off the table altogether, it must make an immediate Armour save. If it succeeds, it will be Pushed Back as far as it can go, and then stop. If it fails, it will be immediately removed as a casualty.

Ongoing Close Combat

Units are not 'locked' in combat in any way as the normal action rules are designed to recreate the fluidity and deadliness of close quarters battle. If a unit wishes to continue a combat, it merely keeps taking Charge actions until it achieves victory, is defeated or chooses to disengage on its turn taking a Move (or Charge) action in a different direction.

However, you will find that once enemy units are within 10" of one another, combat becomes very deadly. The Reaction rules later in this book mean that each time a unit completes an action, all the surviving enemy models nearby may take a Reaction. In practice, the two units will be quickly driven apart or annihilated by the interchange of actions and Reactions.

Cover

Where close combat is concerned, Cover does not grant bonuses to Target and Kill scores in the same way as it does for shooting. However, there are benefits from being concealed in cover when an enemy rushes forward to engage you. See the section on Reactions on page 25 for more details.

READY ACTIONS

A unit taking a Ready action is preparing either themselves or an item of equipment for special use. Some heavy or complicated weapon systems, for example, require a Ready action before they can be fired, while certain types of movement may require Ready actions before they can be used in the next Move action.

In effect, the Ready action is the 'do something special' action.

What a Ready action actually does will be covered by the special rule it is using, usually found in the unit's or weapon's Traits or special rules. Sometimes, a Ready action will be used to actually accomplish something, while others must be performed before using another action (such as when using a Ready weapon or a special movement type). However, there are some rules common to all Ready actions, no matter what they are used for.

A unit performing a Ready action may neither move nor shoot. It stands stationary, performing the Ready action until taking a different action completes it.

You do not have to state what the Ready action is for when the unit makes it (though sometimes it will be obvious). The purpose of the Ready action need only become evident in the following action when you declare it.

A unit may have the ability to perform Ready actions for a variety of different tasks. However, each Ready action may only ever be used for one task, no more.

If a unit performs a Ready action after its first action, the Ready 'status' may be carried over into the next turn (or during a Reaction) if desired.

German infantry use their last action to set up their MG42 LMG

On their turn the American unit fires on the Germans inflicting some casualties... then the Germans react, with the MG42 inflicting heavy losses.

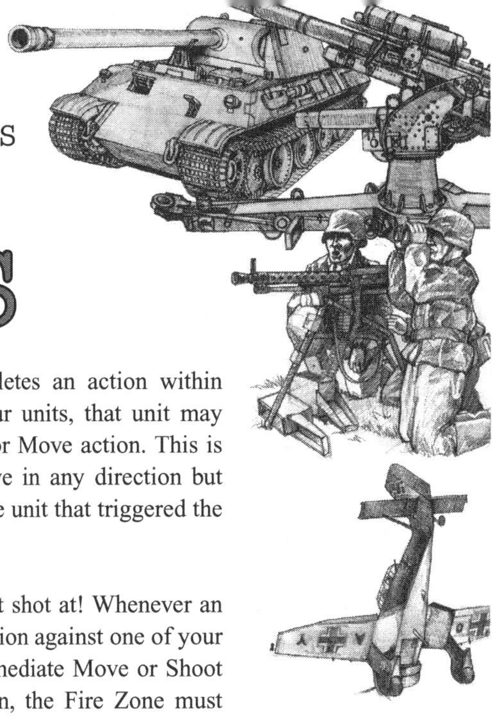

REACTIONS

What makes Battlefield Evolution different from many other games is that units get to react to enemies who get too close. After all, if you saw several enemy soldiers charging towards you, you would not wait around to get beaten up, would you? Also, if you are playing a game, you will not want to hang around for twenty minutes or more while your opponent moves all his models – you want to play as well!

Reactions allow you to take an active part in your opponent's turn, and they may just save a few of your own units from certain death.

Whenever an enemy unit completes an action within 10" of any model in one of your units, that unit may immediately make a free Shoot or Move action. This is called a Reaction. You may move in any direction but any shooting must be aimed at the unit that triggered the Reaction.

You may also react when you get shot at! Whenever an enemy unit completes a Shoot action against one of your units, that unit may make an immediate Move or Shoot action. If a Shoot action is taken, the Fire Zone must include models from the enemy unit that triggered the Reaction.

At the start of their turn the German infantry are already approaching a British Paratrooper position

For their first action the Germans Move forwards toward the Paratroopers

Reactions

A unit may only make one Reaction in every turn, no matter how many times an enemy moves close by or shoots it.

Finally, you can never react to a Reaction!

Ready and Alert Status

Units that took a Ready action as their last action in their player's turn count as being in an Alert Status. They are Readied in the opposing player's turn as well as in their next turn. This means they can potentially use special movement types or Ready weapons during a Reaction.

Reactions Within Cover

Units within Cover have a certain advantage when an enemy charges them. In a defended position, they have the opportunity to flee before the enemy gets too close or gun the attackers down before reaching them.

If a unit is in Cover and is reacting to a Charge action, it may make the Reaction after the target models have moved but before any Close Combat dice are rolled.

Laying Low

A unit does not have to react if it does not want to. This will often happen to units within Reaction range of the enemy who do not wish to call undue attention to themselves or are waiting for another enemy to make a tactical mistake. In real terms the unit might be hiding out as best it can while it awaits reinforcements or retrieval, or locked in an eyeball to eyeball confrontation with an enemy…

The Germans are now within 10” of the Paratroops who may now choose to React in one of two ways - either by Shooting ...

...or Moving

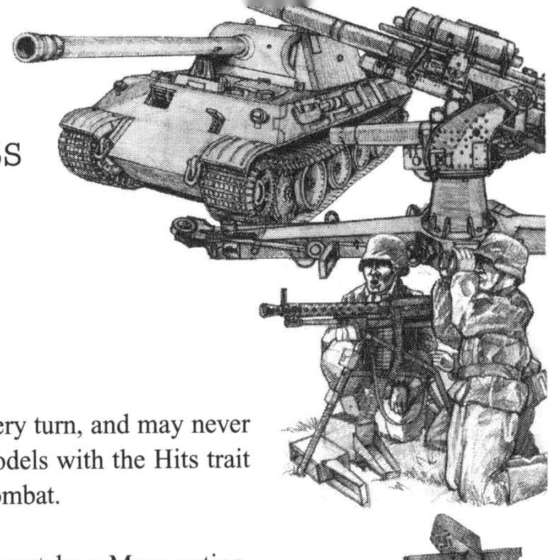

TRAITS

Both models and weapons can have traits that mark them as special. A trait is a special rule that can allow a unit to move in a special way, for example, or make a weapon more powerful.

This chapter lists all the traits used in Battlefield Evolution: World at War, and you should refer to it when coming across a trait in a weapon or unit description later in the book.

AA: This is an anti-aircraft weapon. Only weapons with this trait are capable of affecting Air Units moving at Loiter speed or faster; all others are too limited in their ability to target such rapid enemies.

AA-Only: This weapon may only be used against Air Units that are flying. It may not be used against other units or Air Units that are grounded.

Accurate: Accurate weapons may re-roll any Damage Dice that fail to hit their targets. The Accurate trait is ignored if the weapon is fired as Artillery.

Artillery: The weapon may be fired as Artillery. See page 21 for more details.

Auto: An Auto weapon creates a 6" radius Fire Zone, rather than a 3" one.

Climb/X": This is a special movement type (see page 15). The model uses the special movement distance indicated, counting any impassable and difficult terrain as clear. Models can climb over several actions if required; simply make a note of the height reached or place a dice beside the model as a reminder.

Flame: Flame weapons engulf their targets with a torrent of fire. They ignore any Cover the target may be in and no Dodge save is possible against their effects.

Hits/X: Each time a model with multiple Hits fails its saving throw, it loses one Hit. If the model suffers a Kill result it loses two Hits instead of one. If a model with 6 or more Hits is reduced to just 1, it will become Crippled, and may only take one action every turn, and may never make a Reaction. In addition, models with the Hits trait are never pushed back in close combat.

Immovable: This model may never take a Move action. With a Ready action, it may turn in place to a new facing. If other models in the unit do not have the Immovable trait, they may freely move away from this model at any time, leaving it out of command.

Independent: Independent models have a great deal of flexibility on the battlefield. Independent models are somewhat different as they can function as a one-model unit if they wish *or* join a friendly unit. They can make these choices as they go, so one could potentially join a unit and then leave it later to join another unit or revert to operating as a one-model unit again. Independent models can join a unit's action if they are within its unit leader's command range. They can also form a new unit from any out of command models within command range. The independent model becomes the new unit's leader for the remainder of the game. Note that Independent models are limited to a maximum of two actions per turn as usual.

Lethal Zone/X: This weapon has a large blast effect, affecting multiple models with each shot. See page 18 for more details.

Lumbering: This model is large or slow. It may never make a Reaction.

Multifire/X: Sporting multiple weapon systems, this model can lay down a hail of fire. The Multifire trait shows how many weapons the model can use in a single Shoot action. Any weapons carried may be used. If a model has the Multifire/All trait, it may fire all of its weapon systems in a single Shoot action.

Multihit: High penetration or massive firepower allows this weapon to rip through larger models such as armoured vehicles. Multihit weapons that equal or exceed the Target characteristic of a model score two Hits instead of one. Multihit weapons that equal or exceed the Kill characteristic of a model score four Hits instead of two.

Traits

One-Shot: A One-Shot weapon may only be used in one Shoot action during the entire battle. Once used, it is gone!

Parry/X: A model with the Parry trait or carrying a weapon with the Parry trait can force a single enemy model attacking it to re-roll one of its Close Combat dice during an enemy Charge action. A model may only Parry something up to twice its own Size (that is, a Size 1 model can Parry a Size 2 model but not a Size 6 model). Parry/2 or Parry/3 weapons allow the model to parry two or three Close Combat dice respectively against any opposing models attacking it.

Piercing/X: The model's attacks are naturally lethal and can carve through the heaviest armour with ease. The Armour save roll of models struck by the model's Damage Dice is reduced by the Piercing value. For example an Armour save of 6+ will be impossible to make if struck by a Piercing/1 weapon. If the Trait is possessed by a model, it will affect only its Close Combat attacks. If it is possessed by a weapon, it will affect only the Damage Dice rolled for that weapon.

Prone: Some weapons, such as many man-portable machine guns, can be set up with a bipod or tripod, and so become far more effective in laying down suppressive fire. If a Ready action is taken to go prone and set up a weapon with this trait, it immediately gains the FaF fire arc. However, the number of Damage Dice the weapon has is doubled in all subsequent Shoot actions. These extra Damage Dice will remain until the model moves (the rest of its unit is free to take Move and Charge actions, without negating the model's extra Damage Dice). Another Ready action can be taken later on to go prone and set up the weapon again.

Models of Size 2 or greater may gain the benefit of Prone weapons without taking a Ready action, so long as they do not take a Move or Charge action in the same turn as they shoot.

Ready: A Ready weapon can only be fired by a unit whose preceding action was a Ready action (see page 24).

Retaliate: The Retaliate trait represents volatile ammunition, explosions, crashes and other devastating effects accompanying the models' demise. If a model has the Retaliate trait it and is removed form play due to an enemy rolling its Kill score or higher, then it will roll all of its Close Combat dice against every model (friend or foe) within its Lethal Range. When rolling Damage Dice for Retaliation, only the following traits are still applicable; Flame, Multihit, and Piercing/X.

For weapons use the Damage Dice instead of the Close Combat dice. The Lethal Range of the weapon will be the Lethal Zone of the weapon or the Lethal Range of the model carrying it, whichever is greater.

Slow: Some weapons take time to aim, recharge or require constant reloading. A Slow weapon may only fire once every turn, and may not fire during a Reaction.

Tracks/X": This is a special movement type (see page 15). The model uses the special movement distance indicated and it can always move up to this distance, even if it moves from, over or into difficult terrain (but not impassable terrain).

Wheels/X": This is a special movement type (see page 15). While using this special movement mode a wheeled model moves at much greater speed than normal but is in danger of skidding and losing control if damaged. If a wheeled vehicle is damaged during a game turn in which it has used its special movement, it will be moved D6" away from the source of damage.

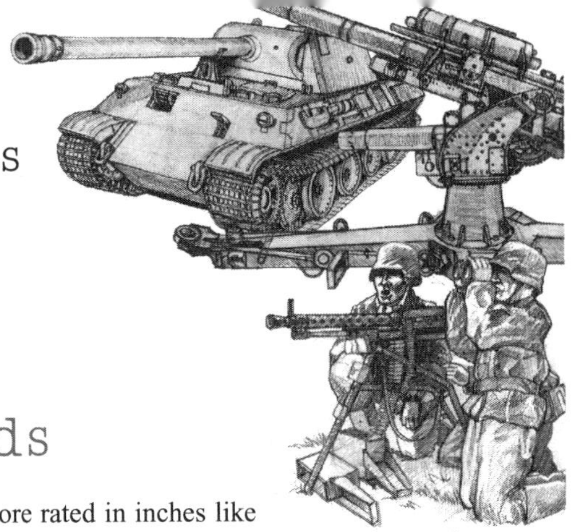

AIR UNITS

The use of air power can fundamentally change the balance between otherwise equal armies and he who holds superiority of the air will have a superb advantage over his enemy. Air Units are expensive in terms of points but, while unable to take ground and thus win battles by themselves, can swing the game in your favour if used properly.

Air combat forms a mini-game in its own right over the battlefield. If either player is using air power, the game gains an additional 'Air Phase' after both players have taken their turns, as shown below.

Game Turn Start
First Player's turn
Second Player's turn
> Air Phase <
Game Turn End

Both players act in the Air Phase, taking turns to move and fight with their Air Units. Each Air Unit gets two actions in the Air Phase, just like ground units do in their turn. Only Air Units may perform actions in the Air Phase, though other units may have the chance to React.

Entering the Battlefield

All Air Units use the Reserves rule (see page 39). As such, they will enter the battlefield in the second turn or later from off-table. An Air Unit that is moving onto the battlefield is placed at any edge of the board at the beginning of the Air Phase. This placement indicates its entry point and heading.

Who Goes First

If both players have Air Units on the tabletop in an Air Phase, the player who deployed his army first will move one Air Unit of his choice first. After the first Air Unit has taken its two actions, the opposing player then takes two actions with one of his Air Units. Both players alternate units in this way until all Air Units have taken their actions.

Flight Speeds

Air Units do not have a Move score rated in inches like other models. Instead, the Move score details the flight speeds that the Air Unit is capable of.

You must decide what flight speed an Air Unit will use at the start of its first action in the Air Phase. It may not alter its flight speed during the rest of its actions in that phase. An Air Unit may not use a flight speed that is not listed in its Move characteristic – it simply cannot fly in a manner contrary to its design.

Players do not have to declare flight speed at the start of the Air Phase – that decision is made when an Air Unit takes its first action.

The possible flight speeds are as follows.

Loiter: A highly manoeuvrable flight state where the Air Unit is being kept aloft primarily by its motive power source rather than its lift surfaces (wings).

Cruise: Medium speed flight, balanced against good manoeuvring ability. Most Air Units capable of long-distance flight use the Cruise flight speed for this purpose but faster aircraft can throttle back to this speed too.

Burn: Full power, throttle to the wall, roaring along with every ounce of power to push the Air Unit forward. Not very manoeuvrable but very, very fast.

Air Target Characteristic

An Air Unit has a Target score to represent its size and sturdiness like any other model. Fast moving targets are harder to hit, however, so an Air Unit always uses the highest of its own Target score and its flight speed's Target score, as shown on the table, except against other air units which only use its Target score.

Flight Speed	Minimum Move per Move Action	Maximum Move per Move Action	Max. Turn Angle	Target	Crash Distance
Loiter	10"	20"	90°	6+	D10"
Cruise	20"	40"	45°	7+	2D10"
Burn	30"	60"	45°	8+	3D10"

Air Actions

During the Air Phase, players alternate taking two actions with their Air Units, until all Air Units have taken two actions and the Air Phase finishes. An Air Unit must always take two actions, unlike other units which do not have this requirement.

All Air Units must take a Move (Fly) action as their first action. What an Air Unit does for its second action is entirely up to you, chosen from the list below.

Unless otherwise stated, all Air Units will ignore all terrain for the purposes of movement. All weapons may only fire into the forward facing, unless otherwise stated.

Charge (Flyby)

Only Air Units that have this listed in their special rules may choose this action. The Air Unit takes a Move (Fly) action as normal. However, if it makes contact with another Air Unit, both roll Close Combat dice. Results are calculated simultaneously. If the charging Air Unit is still airborne, it continues its Charge (Flyby) action, possibly contacting another enemy and causing further simultaneous Close Combat rolls.

Move (Land)

This action may not be performed if the Air Unit is using the Burn flight speed. The Air Unit moves directly forward any distance up to its minimum Move. If, at the end of this movement, there is any impassable terrain within the Air Unit's Size score, it remains an Air Unit. If not, it lands and is now treated as a ground unit (see page 33).

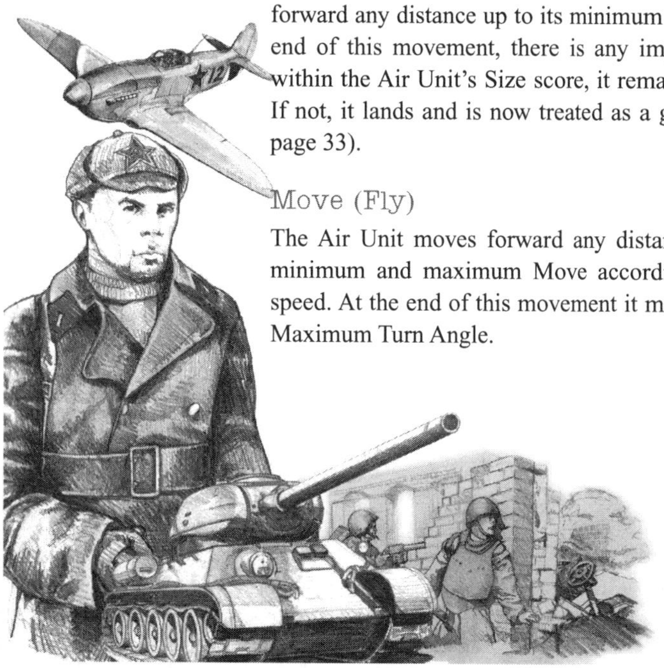

Move (Fly)

The Air Unit moves forward any distance between its minimum and maximum Move according to its flight speed. At the end of this movement it may turn up to its Maximum Turn Angle.

Shoot (X)

Having deliberately slowed its movement to stabilise its attack run, the Air Unit unleashes its firepower upon the enemy, whether this is through bombing, strafing or dogfighting. Note that all weapons fired or deployed by an Air Unit must create Fire Zones in its forward facing, unless specifically stated otherwise in its army list entry.

- *Shoot (Bomb):* The army list entry of an Air Unit will determine how many ordnance munitions may be deployed in a single Shoot (Bomb) action. All ordnance dropped will hit an Impact Point nominated by you anywhere along the flight path the Air Unit moved along in the current turn. Once the Impact Point has been nominated, roll 1D10. Move the Impact Point this many inches towards your Air Unit. Then roll another 1D10 and move the Impact Point this many inches directly away from your Air Unit.
- *Shoot (Dogfight):* The Air Unit may fire its weapons, other than ordnance (munitions classified as dropped weapons), at another Air Unit. The weapons' usual Range scores are used for these attacks. All weapons must create a Fire Zone centred on one enemy Air Unit.
- *Shoot (Strafe):* The Air Unit may fire its weapons, other than ordnance (munitions classified as dropped weapons), at any ground unit. The weapons' usual Range scores are used for these attacks. All weapons must use the same Fire Zone.

Reactions

Air Units may never make Reactions.

Returning to the Battlefield

Air Units can pass completely over the battlefield during their Air Phases and, because of their speed, can fly off

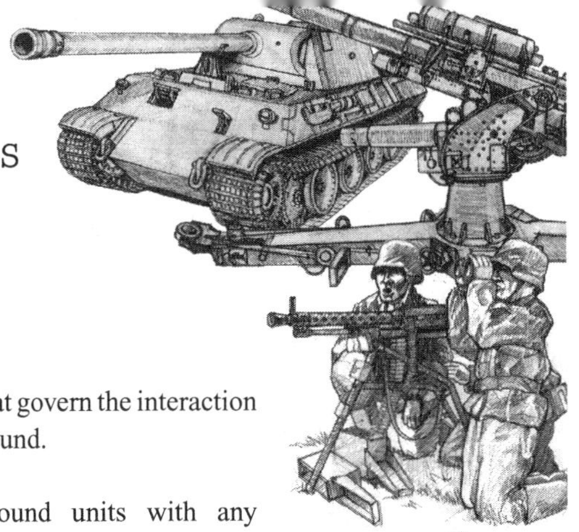

a table edge. If an Air Unit leaves the table, it may try to turn about and head back to the battlefield.

Unfortunately events in the wider world can prevent this – it might run into anti-aircraft fire or enemy aircraft, get called on to help somewhere else or decide it is too badly damaged to continue the mission. To represent these conditions an Air Unit that has left the table may roll a D6 at the beginning of each subsequent Air Phase. On a 4 or more, it will re-enter the battlefield. If the roll is failed, it can try again at the beginning of each subsequent Air Phase.

When returning to the table, Air Units can come on from any table edge with any initial heading. You can choose to forgo this roll if you wish, in order to attempt to bring the Air Unit back in a later turn.

If several Air Units are entering or re-entering the battlefield in the same Air Phase, players alternate placing their models at the beginning of the phase, with the player who deployed his army first placing the first Air Unit.

Air/Ground Interaction

There are a number of key rules that govern the interaction of Air Units with those on the ground.

- Air Units can attack ground units with any weapons.
- Air Units that land during the Air Phase become ground units until such time as they take off again (see page 33).
- Ground units can only attack airborne Air Units with AA weapons.
- Ground units can react against Air Units that complete an action within 20" of them (rather than the usual 10"). However, they may only choose Shoot as a reaction.
- When determining Line of Sight to or from Air Units, assume they are 12" above the table.

A British Typhoon enters the battlefield

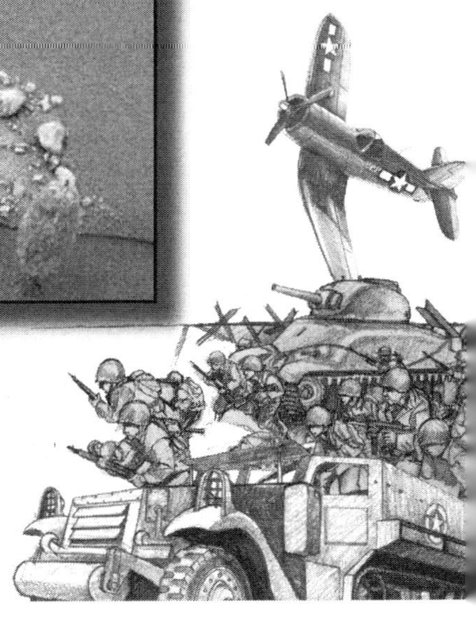

The Typhoon moves forward and engages the Focke Wulf in a dogfight, destroying it.

- Air units ignore Fire Zones not specifically targeted at them (thus ignoring Fire Zones created on the ground).
- By the same token, units on the ground ignore Fire Zones created by an Air Unit attacking another Air Unit.

Destroyed Air Units

Unlike normal models, Air Units are not simply removed from play when they are destroyed – they crash to the ground, and woe betide anything that gets caught underneath. . .

When an Air Unit is destroyed, consult its most recent flight speed on the table on page 30, and roll the Crash Distance dice listed. The Air Unit must move directly forwards *precisely* this distance. Every model and structure then within the Air Unit's Size score in inches will immediately suffer all of the Air Unit's Close Combat dice. The destroyed Air Unit is then removed as a casualty if it is Size 2 or less, or left on the table if it is larger.

Air Units that have landed and are subsequently destroyed while on the ground do not crash and are simply left on the table if Size 3 or greater, or removed if smaller.

Air Transport

Some Air Units are capable of carrying other units for deployment onto the battlefield. This will be noted in their army list entries.

In order to do so, the Air Unit must carry them onto the table as Reserves, and land on the battlefield during the Air Phase. Units within an Air Unit that has landed must use a Move action to disembark but may take no other actions in that turn.

To enter an Air Unit capable of transport, a unit must move on to it, using a Move action. The unti may then be removed from the table, as it has boarded the aircraft. Untis removed in this manner do not count as casualties; they are assumed to be within the Air Unit.

Any units that are within an Air Unit that is destroyed are instantly counted as casualties.

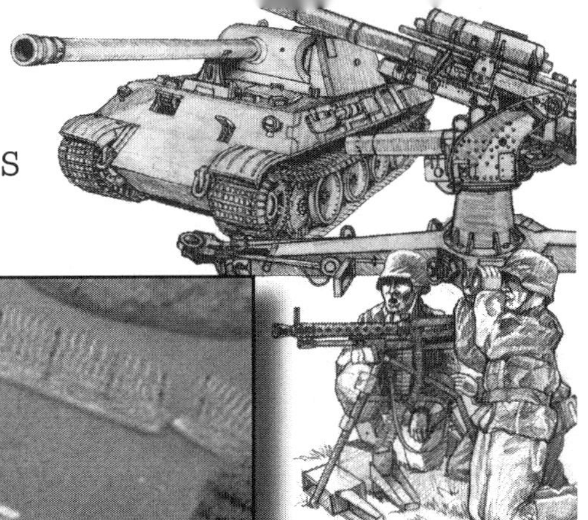

The typhoon then engages the Panzer IV, destroying it, but is in turn shot down by the German anti-aircraft gun when it Reacts.

Landed Units

Air Units are significantly impaired when they have landed. Landed Air units count as ground units in all respects and no longer take actions in the Air Phase. The only actions they may perform are Move (Take Off) or Shoot (Strafe), though only weapons listed in their army list entry as being capable of shooting while on the ground may be used.

Air Units that have a special movement trait become normal ground units when landed.

Move (Take Off): This action requires that the Air Unit's previous action was Ready. Choose a flight speed it is capable of and check to see if any impassable terrain is blocking the Air Unit within its minimum Move for that flight speed. If it is blocked, the Air Unit may be turned to face any direction. The Air Unit then moves *precisely* its minimum Movement directly forwards and is now considered a proper Air Unit. It does not perform any actions in the Air Phase at the end of this Game Turn, however.

STRUCTURES

Battles are often dictated by key structures – forts, bridges, cities, factories, harbours and the like. Military forces use defensive emplacements to help secure areas against threats, while many commanders are forced to engage enemies within an urban environment, a great equaliser of armies where cunning and local knowledge can make up for an imbalance of technology.

A structure is any building placed on the battlefield and is a type of terrain. You should treat them as such when placing terrain on the table and, in fact, you can set up quite involved settlements or parts of cities on your table to create a very exciting battlefield.

All structures have the following scores.

Size: The Size score of a structure is equal to the number of inches (rounded up) measured on its longest side, from the lowest point on one corner, to the highest point on the opposite one.

Target and Kill: These scores work in the same way as for units with multiple Hits.

Hits: This is how many Hits a structure can withstand before being destroyed, and it works in the same way as the Hits traits for models. Multiply the structure's Size score by 1.5 (rounding up) to see how many Hits it can withstand.

Moving Inside a Structure

A structure can only be entered and exited by openings already in it – doors, windows, ramps and so on. A model cannot fit through an entryway that is smaller in inches (across its longest diagonal) than its Size score. Any number of models can use the same the entryway providing they can fit through it and have sufficient movement to do so. Models inside a structure are said to have Occupied it.

If a model is within a multi-level structure it may move up or down one level with a single Move action (with no horizontal movement).

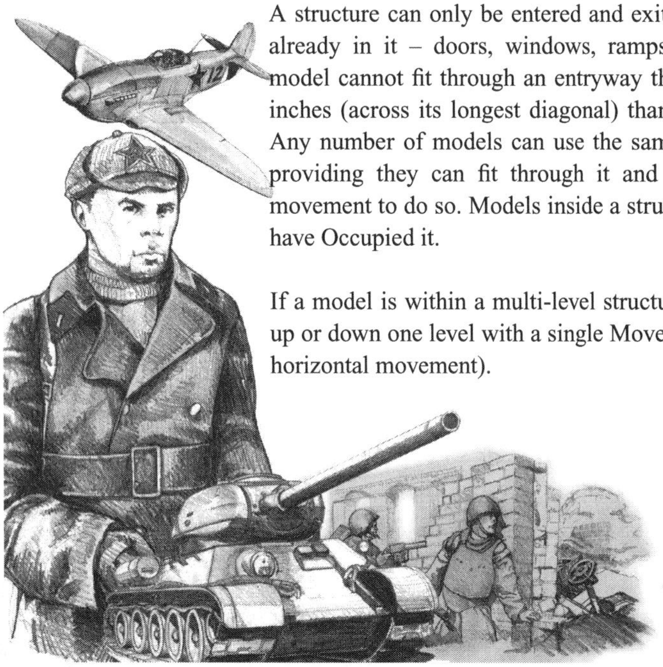

Fighting From a Structure

Each window, door or other opening on a structure is considered to be a Firing Point. Each Firing Point allows one model inside to make a shooting attack against enemies outside the structure. Any model can use any Firing Point that is on the same level of the structure.

A model within a structure cannot have a Line of Sight drawn to it from the outside, nor may it be attacked by Artillery weapons, unless it uses a Firing Point to make an attack itself.

If a model inside a structure uses a Firing Point, it may have a Line of Sight drawn to it for the next full turn (use the Firing Point itself to work out Line of Sight), but will gain the effects of Cover (normally Medium Cover for light buildings, though military and heavy commercial style buildings will usually grant Heavy Cover).

Fighting Within a Structure

If a structure is Occupied by an enemy unit, a Charge action is needed to enter it. The defending models count as being in Cover (see page 34). After any Reactions have been resolved, one model on each side must fight in close combat for every Firing Point. Defending models may use a shooting attack (so long as their weapons do not have the Slow trait) instead of their Close Combat dice for this.

If the attacking models destroy more models in close combat than they lose during this close combat, they move inside the structure (the rest of their unit may follow using subsequent actions). After this, immediately resolve the fight between as simultaneous Charge or Shoot actions by both units. All models may use either Close Combat dice or shooting weapons as they wish. The fighting is close and furious so no more actions take place until the combat is completely resolved – push back rules are not used. Keep rolling for close combat or shooting damage until one unit or the other has been wiped out.

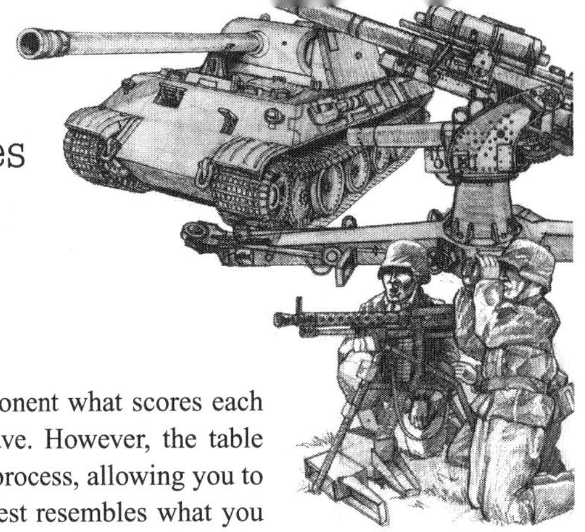

As you can see, fighting in built-up areas is vicious and nasty!

If the attackers do not destroy more models in the initial close combat, then they have been barred from entry and the close combat ends immediately (though they are free to perform another Charge against the structure in their next action).

Attacking a Structure

Shooting and close combat attacks will not harm a structure unless it is specifically declared as the target – otherwise, models are assumed to be trying to shoot through openings at what is inside rather than blast the structure itself. If the structure is declared as the target, other models within the Fire Zone will not be affected by the attack (unless, of course, the structure is destroyed).

Artillery fire (see page 21) and Ordnance weapons from Air Units (see page 16) that place Impact Points on structures will only attack the structure itself, not models inside.

Structure Collapse

If a structure is destroyed, it collapses. Anything inside or near a structure when it collapses is in deep trouble. The collapsing structure rolls a Damage Dice of D10 against every model inside or within its Size in inches. A +1 bonus is added to these Damage Dice for every level (storey) the building has beyond the first (ground level).

You can replace the structure with a ruined version or otherwise scatter some rubble around it.

Example Structures

You should agree with your opponent what scores each structure on the table should have. However, the table below should make this an easy process, allowing you to select type of structure that closest resembles what you have on the table, and simply measuring it to obtain its Size.

Structure Type	Target	Kill
Primitive Building	6+	8+
Housing Block	7+	10+
Commercial Block	8+	11+
Industrial Block	9+	12+
Military Building	10+	13+
Bunker	10+	14+

Very Large Structures

Some structures can be very large, representing huge military complexes or civilian towers. Where it is not practical to split these into several separate structures, use the following rule.

For every multiple of 8 a structure's Size is, add +1 to both its Target and Skill scores, to a maximum of 14+.

Walls

A structure that cannot be entered by models is deemed to be a wall (though in reality it may not be – a large chemical container, for example, could not be entered, but it would use these same rules).

Walls use the same rules as structures for determining their Target, Armour and Kills scores, as well as for being attacked, but no models can enter them at any time. In all other respects, they are counted as terrain (usually providing Medium Cover, though fortified walls will provide Heavy Cover). They are also counted as impassable terrain, except by any models that are able to move onto their ramparts, either by a special movement trait or by an access way, such as stairs.

This very large building is Size ten, so has fifteen Hits.

With a much smaller Size of five this bunker has just eight Hits.

ADVANCED RULES

The rules in this section are not necessary to play the game and can be added when you are comfortable with the previous sections. They will add greater depth to your battles and provide you with more tactical options for your army. Many of the scenarios found later in this book also use the advanced rules detailed here.

Covered Advance

A typical playing surface for any miniatures game is flat with various terrain pieces placed upon it. In the real world, however, very few battlefields are like this. The ground tends to be uneven, filled with patches of vegetation or rock, all of which can be used to gain a little protection from enemy fire.

To represent this, any unit of Size 1 models may make a Covered Advance by performing a Ready action before a Move action. If attacked by an enemy, they will count as being in an obscured Line of Sight, even if they would normally be in a clear Line of Sight.

This only applies to enemy shooting attacks made in the enemy's turn. Models do not receive the benefit of obscured Line of Sight if the enemy is shooting at them because of a Reaction, as they are able to catch the unit as it moves between concealed positions.

Elevated Firing Positions

Infantry nested within terrain or structures can be incredibly difficult to shift without a costly frontal assault or sustained bombardment. However, there are other reasons for infantry to take terrain and dominate the battlefield. Troops that gain a good height advantage over their enemies can often target weak areas of armoured vehicles and similar units, making them extremely vulnerable.

If a unit is more than 6" above an enemy it makes a shooting attack against, and is within 15" on a horizontal level, it is said to be in an Elevated Firing Position.

If the enemy unit has the Lumbering trait, then it will suffer an additional –1 penalty to its Armour rolls against shooting attacks made by units in the Elevated Firing Position.

Emplacements

Emplacements are fortified structures that may be placed within your own deployment zone and used for defensive purposes. They represent anything from small foxholes to large bunker complexes.

Every army has access to the same Emplacements list. You may not choose an Emplacement of a larger Size than you purchase on the table below.

Minefields

Mines are a valid tactical asset for all armies in Battlefield Evolution: World at War. They are used primarily to keep an enemy off balance or make certain areas of territory extremely hazardous. When used in great numbers, mines can be used to completely deny a safe approach to critical locations.

An army may purchase Minefields as if they were Emplacements. Divide the table into 12" by 12" squares, You may purchase one Minefield for each of these squares but may not purchase a square that is within your opponent's deployment zone. These squares must be chosen and noted down before any models are deployed on the table.

Emplacements

Structure Type	Max. Size	Target	Armour	Kill	Hits	Value
Foxhole	2	8+	3+	12+	3	25
Small Bunker	4	10+	2+	14+	6	75
Large Bunker	6	10+	2+	14+	9	150
Fortified Military Structure	10	10+	2+	14+	15	250

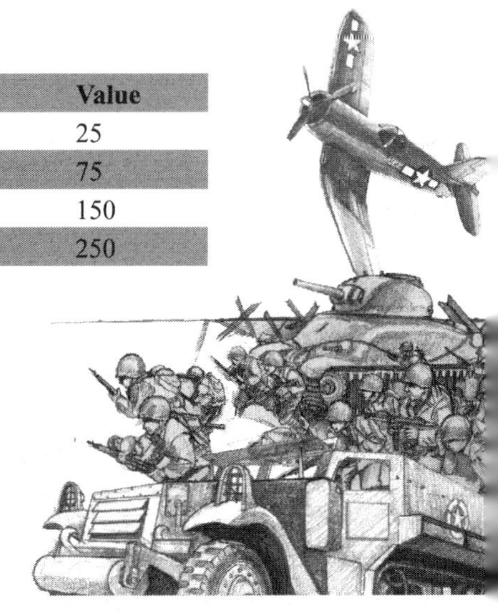

If you purchase a square and fill it with anti-personnel mines, this will cost 150 points.

If you purchase a square and fill it with anti-vehicle mines, this will cost 250 points.

You may purchase both anti-personnel and anti-vehicle mines for the same square at a combined cost of 350 points.

Anti–Personnel Mines

Whenever a unit with Size 1 models completes a Move action within a square filled with anti-personnel mines, roll a D6. On a 5 or more, a mine explodes – roll Damage Dice 5xD6 against the unit, allocating dice to the models closest to the enemy deployment zone first. Only one model in the unit has to enter the minefield to trigger this, and the mines will still effect the entire unit.

Once an anti-personnel minefield has been located in this way, a unit with Size 1 models will only explode a mine on a roll of 6 more, if it takes a Ready action first.

Anti-Vehicle Mines

You may choose whether your minefield will detonate when a unit containing Size 2 or greater models completes a Move action within it, or whether it will only detonate when a unit containing Size 3 models completes a Move action within it. This decision must be made before you choose which square the minefield will lie within.

Whenever an appropriately sized model completes a Move action within a square filled with anti-vehicle mines, roll a D6. On a 4 or more, a mine explodes – roll Damage Dice 2xD10+3 with the Piercing/4 trait against the model.

If a vehicle moves into a minefield and out again in a single Move action, the D6 must be rolled at the edge of the minefield, just before it leaves.

Night Fighting

Many fights and battles in World War II took place during the night. The following rules may be used in Engagements to represent these conditions.

- All weapon ranges are halved.
- All Target and Kill scores get a +1 bonus during shoot actions. This bonus is cumulative with bonuses from Cover or obscured Line of Sight, so most basic weapons can't kill an enemy outright during a Night Fight.
- The radius of all Fire Zones (including those with the Auto trait) is reduced to 3".

Parachute Deployment

Paratroopers trained to use slow flying transport aircraft to enter battle are usually trained to disembark in short order during flight. Though the troops risk coming under fire from nearby enemy forces, this tactic means the troops can be deployed nearly anywhere.

A Parachute Deployment uses the Reserves Rule below - however you are not holding back a single unit, you are holding back your entire force! If you want to deploy your whole force via Parachute Deployment, you have to do so in a single turn.

Whenever a player deploys a unit by parachute, he places the unit leader anywhere on the tabletop.

To represent deviation by the wind blowing during the jump, the opponent rolls 4D6. He may move the unit leader as many inches in any direction as the total of the dice. All following parachuting units have to be moved into the same direction. The opponent only rolls the distance for these units.

The parachuting player then rolls 2D6 and moves the unit leader in any direction up to a number of inches equal the dice roll, representing the control his troops have over their parachutes. The rest of the unit members are then deployed within 6" of the unit leader.

The opposing player rolls 1D6 Damage Dice against every model that is lands within Cover or on top of structures.

If a unit leader is moved off the table, his unit become a Reserve unit that comes into play from the table edge he was blown off.

The Parachute Deployment is considered to be the first action of the deployed unit (a Ready action), so every

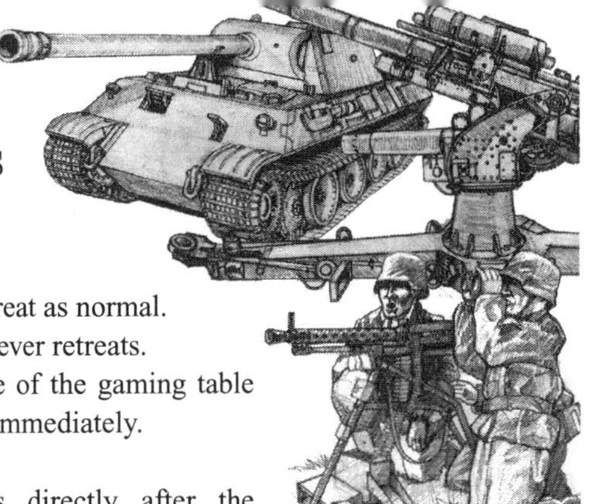

enemy unit within 10" can react to them! The unit's second action can then be performed as desired.

Reserves

Reserves are units held back from an initial advance. Tactically speaking, they allow a commander to assess the power of an enemy before committing his full force to battle and also allow greater flexibility in unit placement to attack at the weakest point.

Units using the Reserve rules can arrive on the table during your second player turn. You may delay a reserved unit's arrival to a later turn if you wish.

Reserves can move on from any table edge in your deployment area, ignoring the proximity of enemy units (unless otherwise specified).

Models with special movement traits (such as Wheels or Tracks) may employ them to enter play from Reserves, counting as having been Readied prior to their first action. This allows them to come on to the table, all guns blazing!

Reserve units must come on the table before the last turn of the game, or they are considered lost for the purposes of victory – you can assume they were engaged by roving enemy forces or perhaps they simply got lost. In scenarios with no fixed turn length, you must judge when to bring on Reserves very carefully, or risk losing them altogether!

Retreat

As soon as units take casualties from shooting attacks, they have a tendency to retreat from combat. It is one thing to just keep your head down when the bullets starts to fly, but to see a comrade die from enemy fire is something else. Whenever a unit has to remove a model from the battlefield because of an enemy Shoot action, it has to retreat a short distance.

The unit has to move 1 inch for each model it has removed because of the opponents shooting action. The unit must move as directly as possible away from the source of damage. The only exceptions are listed below.

- A unit in Cover does not have to Retreat, but may do so if the player chooses. Note that units that are only Obscured must still Retreat as normal.
- A unit with the Hits/X trait never retreats.
- A unit that reaches any edge of the gaming table stops the Retreat movement immediately.

The Retreat movement happens directly after the opponent's Shoot action before any Reactions are taken. A unit may be forced to Retreat several times per game turn if it comes under fire multiple times.

Suppression

Another important factor in warfare is Suppression. When the bullets start flying, even the most hardened troops have a tendency to keep their heads down!

If every model in a unit is allocated a Damage Dice in a single Shoot action, then every Size 1 model will immediately lose its next action, and may only Move as a Reaction. If this happens more than once during a single turn, it will lose a total of two actions, though may still React by moving. A unit may never lose more than two actions from Suppression.

Out of command models may be ignored for the purposes of allocating Damage Dice to work out Suppression – in effect, you need only worry about allocating Damage Dice to models that are still in command in order to suppress them. However, you may choose to Suppress out of command models if you wish, though a model may only ever lose a maximum of two actions, even if both Suppressed and out of command.

Suppression can only affect Size 1 models – larger models simply shrug off the effects.

Tactical Withdrawals

Occasionally, you may have to concede defeat. For all your efforts, the enemy has out-fought you every step of the way and there is little choice left but to die or retreat. In many games, you may find it makes little difference whether your army lives or dies but for certain scenarios and when playing campaigns, getting your men out of the line of fire can make a huge difference.

If you intentionally move a unit off the table, you immediately cede a minimum 25% of its value to your opponent in Victory Points. If that unit would normally earn your opponent more

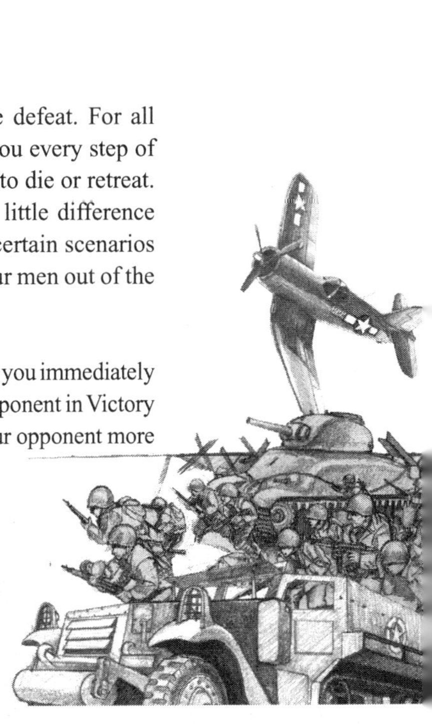

Victory Points, perhaps because of scenario conditions or because the unit has been reduced to below half strength, then your opponent earns the higher amount for it.

This rule does not apply to Air Units, which will spend a great deal of time off the table anyway.

Transport Vehicles

Some vehicles are capable of carrying other units into battle. Normally, these will be squads of infantry, but some very large vehicles can carry smaller vehicles inside them as well.

Vehicles capable of acting as transports will have the Transport special rule in their army list descriptions. This entry will note how many Size points of models the vehicle can carry, and what the maximum Size score of each of these models may be.

Models may enter a transport vehicle simply by moving on to it, and may leave with a Move action in the same way. If models either enter or leave the transport vehicle, then both those models and the transport vehicle may only take a single action in that turn.

A unit may deploy inside a transport vehicle purchased for it in the army list, or it may be deployed seperately, with the transport vehicle counting as a seperate unit.

If the transport vehicle is destroyed, every model on board will be immediately removed as a casualty.

Vehicles

Vehicles (units with the Tracks, or Wheels traits) will often have different Target, Save and Kill scores in their army list description for each facing, since many vehicles have weaker armour to the sides and rear that enemies can take advantage of. If a vehicle is attacked, use the appropriate Target, Save and Skill scores, depending on which facing the enemy is within.

Shooting with Vehicles

Nearly all vehicles in World War II had un-stabilised guns. Tank crews were trained to stop before shooting to overcome this disadvantage.

Any weapon mounted on a vehicle must re-roll all successful attacks (any Damage Dice that roll equal or above an enemy's Target score) if the vehicle performs a Move action in the same turn, even if the Move action uses a special movement mode (see page 15).

Victory Points

Many of the scenarios included later in this book use Victory Points to finely grade the level of victory. Often in battle, simply smashing the enemy is not sufficient to gain total victory – instead, you will have a mission to perform and it is your success at this that will be measured, as much or even more than your ability to destroy the enemy.

When Victory Points are used to find out who wins a battle, you will earn them in the following ways.

- You will gain Victory Points equal to the Value of every unit you completely destroy.
- If you do not completely destroy a unit, you will gain Victory Points equal to half the Value of the unit if you destroy at least half the models within it.
- If a unit comprises a single model and you do not completely destroy it, you will gain Victory Points equal to half its Value if you reduce its Hits below half its starting total.
- A destroyed unit will always yield its full Victory Points, regardless of whether any One-Shot weapons or other limited forms of equipment have been used or not.
- You may also have a chance to gain bonus Victory Points – these will be noted in the scenario you play.

LATE WAR BRITISH ARMY INFANTRY

During the Second World War, the British Army still policed a huge Empire and had its strength spread across the globe. Initial conflicts were met with disaster in France, Africa and the far east, but during the latter years of the war, the British Army had a new confidence, bolstered by victories at El Alamein, in Italy and the D-Day invasion of Normandy.

Platoon Structure

The standard section size for most British Army platoons is ten soldiers, consisting of two teams. The section is led by a Corporal, while the second team is led by a Lance Corporal.

Three such sections make up a standard platoon, led by a Lieutenant, who is in turn assisted by a Sergeant. Three platoons make a rifle (infantry) company.

A platoon that has been on operational duty for some time may vary wildly from this, however. Heavy losses can result in fewer and smaller sections, while a platoon involved in heavy combat may become 'reinforced' with more soldiers, giving it more sections.

You can make any platoon with the British Army list, from one that has just been deployed, to a grizzled set of veterans that have taken heavy losses or been reinforced.

Each unit in a platoon is considered to be an Asset of some type. You may include the following Assets in your platoon, so long as you obey the minimum and maximum units listed for each. You may take one platoon for every 1,500 points in your force.

0-1 Command Assets
2-4 Squad Assets
0-1 Support Asset per Squad Asset
0-3 Armour Assets
0-2 Air Assets

For details on more Support, Armour and Air Assets see the British Vehicle and Air Assets List on page 61).

Special Rules

The following special rules apply to all British Army Infantry platoons.

Carrier Platoons

The specialised Carrier Platoons were designed to move infantry under armoured protection, becoming what would later be called mechanised infantry.

A British Army Infantry platoon may choose to use either Rifle Sections or Carrier Sections – you may not take both in the same platoon, though different platoons in the same force may have either.

Command Assets

The following are all the Command Assets available to a British Army Infantry platoon.

Command Section - 85 points

Trained to use their own initiative, British Army officers are able to quickly cope with changing battlefield conditions.

Type	Size	Move	Close Combat	Target	Save	Kill	Traits
Lieutenant	1	4"	D6	4+	6+	6+	-
Sergeant	1	4"	D6	4+	6+	6+	-
Corporal	1	4"	D6	4+	6+	6+	-
Soldier	1	4"	D6	4+	6+	6+	-

Team 1: Lieutenant (unit leader) with a Sten SMG, Sergeant with a Sten SMG, Soldier with a Lee-Enfield rifle.

Team 2: Corporal with a PIAT, Soldier with a Lee-Enfield rifle.

Teams: The Corporal may act as unit leader to Team 2, splitting off to act as a separate unit.

Options: Drop Team 2 for –45 points. Add one additional Team (same composition as Team 2) for +45 points. The section may be mounted in a Bedford Truck for +50 points, or up to 2 Bren Carriers for +70 points each, or an M3 halftrack for +80 points.

Supporting Artillery: This unit may purchase up to 3 uses of Supporting Artillery. Calling in Supporting Artillery requires a Ready action. This is used as an Artillery attack, but the initial Impact Point must be in the unit's Line of Sight. There is no minimum range for this attack (yes, you can call it down right on your own head!). Light Artillery costs +50 points per use, while Heavy Artillery costs +100 points per use.

Squad Assets

The following are all the Squad Assets available to a British Army Infantry force.

Rifle Section - 140 points

The backbone of the British Army, the steady rifleman was suitable for most battlefield tasks and could always be relied upon.

Type	Size	Move	Close Combat	Target	Save	Kill	Traits
Corporal	1	4"	D6	4+	6+	6+	-
Lance Corporal	1	4"	D6	4+	6+	6+	-
Soldier	1	4"	D6	4+	6+	6+	-

Team 1: Corporal (unit leader) with a Sten SMG, 6 Soldiers with Lee-Enfield rifles.

Team 2: Lance Corporal with a Sten SMG, Soldier with a Lee-Enfield rifle, Soldier with a Bren LMG.

Teams: The Lance Corporal may act as unit leader to Team 2, splitting off to act as a separate unit.

Tank Assault: Infantry in this period used a variety of grenades or mines to assault armoured vehicles. All members of the unit are assumed to carry Anti-Tank Explosives. These may be placed on any model of Size 2 or greater with a Charge if the target ends up within the Lethal Range of the charging unit. They will explode at the end of the charging unit's next turn.

Options: The section may be mounted in a Bedford Truck for +50 points, or an M3 halftrack for +80 points.

Carrier Section – 140 points

The Carrier Section was a unique formation to perform re-supply, reconnaissance, and support missions or act as mobile strike force.

Type	Size	Move	Close Combat	Target	Save	Kill	Traits
Corporal	1	4"	D6	4+	6+	6+	-
Lance Corporal	1	4"	D6	4+	6+	6+	-
Soldier	1	4"	D6	4+	6+	6+	-

Team 1: Corporal (unit leader) with a Sten SMG, 3 Soldiers with Lee-Enfield rifles.

Team 2: Lance Corporal with a Sten SMG, 3 Soldiers with Lee-Enfield rifles.

Team 3: Lance Corporal with a Sten SMG, 3 Soldiers with Lee-Enfield rifles.

Teams: The Lance Corporals may act as unit leader to Teams 2 and 3, splitting off to act as a separate unit.

Tank Assault: Infantry in this period used a variety of grenades or mines to assault armoured vehicles. All members of the unit are assumed to carry Anti-Tank Explosives. These may be placed on any model of Size 2 or greater with a Charge action if the target ends up within the Lethal Range of the charging unit. They will explode at the end of the charging unit's next turn.

Options: One Soldier from Teams 2 and 3 may swap his Lee-Enfield rifle for a Bren LMG for +30 points each. Each Team may have an Universal Carrier for +70 points each.

Support Assets

The following are all the Support Assets available to a British Army Infantry force.

Anti-Tank Team - 45 points

Armed with the PIAT, this team was trained in the dangerous task of hunting down and eliminating enemy armour.

Type	Size	Move	Close Combat	Target	Save	Kill	Traits
Soldier	1	4”	D6	4+	6+	6+	-

Unit: Soldier (unit leader) with a PIAT, Soldier with a Lee-Enfield rifle.

Bren Team - 55 points

Laying down a hail of fire to support the rest of the platoon’s advance, this small team has firepower that outweighs their size.

Type	Size	Move	Close Combat	Target	Save	Kill	Traits
Soldier	1	4”	D6	4+	6+	6+	-

Unit: Soldier (unit leader) with a Bren LMG, Soldier with a Lee-Enfield rifle.

Sniper Team - 60 points

Using cover and gaining access to advantageous positions on the battlefield, this small team can stall entire squads of enemy soldiers.

Type	Size	Move	Close Combat	Target	Save	Kill	Traits
Soldier	1	4”	D6	4+	6+	6+	-

Unit: Soldier (unit leader) with a Scoped Lee-Enfield rifle, Soldier with a Scoped Lee-Enfield rifle.

Stealthy: If a model from this unit is in Cover and did not make a Shoot action or reaction in the last turn, enemy models more than 20” away may not draw Line of Sight to it.

Ambush: This unit may be concealed in an area of terrain outside the enemy’s deployment zone before the game begins. Note the location of the unit before any models are deployed on the table. You can reveal the ambush by placing the section on the table at any time after a unit (friendly or otherwise) has completed an action. However, the unit must be revealed by the end of turn three or be counted as destroyed.

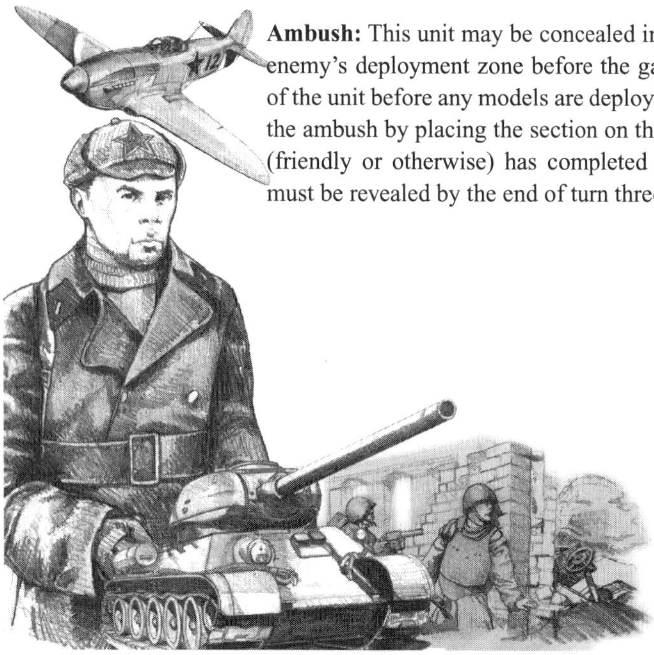

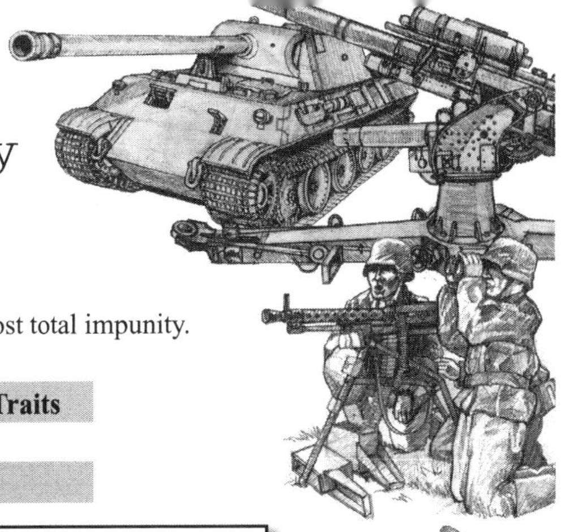

Medium Mortar Team – 75 points

Raining death from the other side of a building or hill, this team can attack the enemy with almost total impunity.

Type	Size	Move	Close Combat	Target	Save	Kill	Traits
Corporal	1	4"	D6	4+	6+	6+	-
Soldier	1	4"	D6	4+	6+	6+	-

Unit: Corporal (unit leader) with a Lee-Enfield rifle, Soldier with a 3" Mortar (FaF), Soldier with a Lee-Enfield rifle.

Options: The Corporal may swap his Lee-Enfield rifle for a Sten SMG for +0 points.

MMG Support Team – 75 points

With a Vickers medium machine gun to support an assault, there are few enemies who can stand up to such a determined attack.

Type	Size	Move	Close Combat	Target	Save	Kill	Traits
Corporal	1	4"	D6	4+	6+	6+	-
Soldier	1	4"	D6	4+	6+	6+	-

Unit: Corporal (unit leader) with a Lee-Enfield rifle, Soldier with a Vickers MMG, Soldier with a Lee-Enfield rifle.

Options: The Corporal may swap his Lee-Enfield rifle for a Sten SMG for +0 points.

6 pdr Anti-Tank Gun – 75 points

Able to destroy enemy armour at a distance, the six pounder is best used in an ambush or when on the defensive.

Type	Size	Move	Close Combat	Target	Save	Kill	Traits
Sergeant	1	4"	D6	4+	6+/6+	6+	-
Soldier	1	4"	D6	4+	6+	6+	-

Unit: Sergeant (unit leader) with a Sten SMG, Soldier with a 6 pdr Anti-Tank Gun (FaF), Soldier with a Lee-Enfield rifle.

Crew Weapon: The 6 pdr Anti-Tank Gun can only shoot if at least one crew member is in base contact with it.

Cumbersome: A Ready action must be taken before the 6 pdr Anti-Tank Gun may be moved. The Move score of all models in the unit depends on the number of models still manning the weapon.

3 crew = 3" maximum move,
2 crew = 2" maximum move and
1 crew = No movement possible.

If the crew choose to abandon the 6 pdr Anti-Tank Gun, they may move normally.

Gun Shield: The Sergeant and Soldiers count as being in Medium Cover when shot from the Forward facing.

Options: This unit may have a Bedford Truck for +50 points or an M3 halftrack for +80 points (see US vehicle and Air Assets list).

Engineer Section – 140 points

Roaming the battlefield, the Engineers provide ad hoc support in close assaults or against armoured vehicles.

Type	Size	Move	Close Combat	Target	Save	Kill	Traits
Sergeant	1	5”	D6+1	4+	6+/	6+	-
Corporal	1	5”	D6+1	4+	6+	6+	-
Soldier	1	5”	D6+1	4+	6+	6+	-

Team 1: Sergeant (unit leader) with a Lee-Enfield rifle, 2 Soldiers with Lee-Enfield rifles.

Team 2: Corporal with a Flamethrower, 2 Soldiers with Lee-Enfield rifles.

Elite Training: The unit gains a +1 bonus to their Armour Saves when in Cover. They may also re-roll their Close Combat dice.

Fearless: The unit discards the first Suppression dice it suffers every turn.

Fire Discipline: The unit need only discard one Damage Dice for each extra Fire Zone created.

Teams: The Corporal may act as unit leader to Team 2, splitting off to act as a separate unit.

Tank Assault: Infantry in this period used a variety of grenades or mines to assault armoured vehicles. All members of the unit are assumed to carry Anti-Tank Explosives. These may be placed on any model of Size 2 or greater with a Charge action if the target ends up within the Lethal Range of the charging unit. They will explode at the end of the charging unit’s next turn.

Options: Any member of the unit may swap their Lee-Enfield rifle for a Sten SMG for +0 points. The whole unit may be equipped with Improved Anti-Tank Explosives for +30 points. An Engineer Section may have 2 Universal Carriers for +140 points, or a Bedford Truck for +50 points.

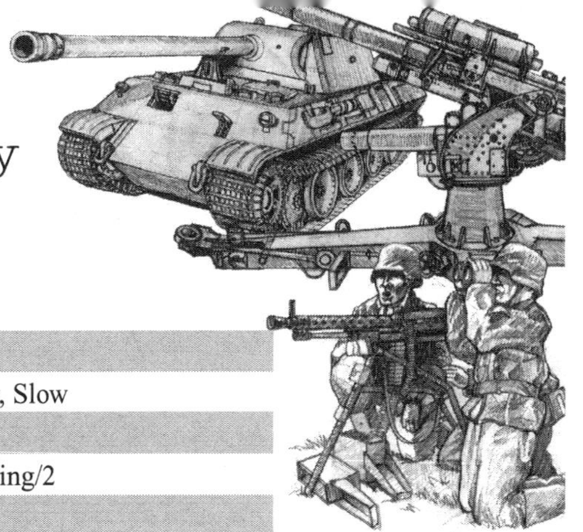

Late War British Armoury

Weapon	Range	Damage	Traits
3“ Mortar	48”	D6	Artillery, Lethal Zone/3”, Ready, Slow
6 pdr Anti-Tank Gun	36”	D10+2	Lethal Zone/1”, Ready, Slow
Anti-Tank Explosives	Placed	D10	Lethal Zone/2“, One-Shot, Piercing/2
Bren LMG	24”	2xD6	Auto, Prone
Flamethrower	10”	3xD6	Flame, Slow
Heavy Artillery	Special	2xD6+1	Lethal Zone/4”, One-Shot
Improved Anti-Tank Explosives	Placed	2xD10+2	Lethal Zone/2”, One-Shot, Piercing/3
Lee-Enfield Rifle	24”	D6	-
Light Artillery	Special	D6	Lethal Zone/4”, One-Shot
PIAT	18”	D6+3	Lethal Zone/ 1”, Piercing/1
Scoped Lee-Enfield	36”	D6	Accurate
Sten SMG	12”	3xD6	-
Vickers MMG	48”	2xD6+1	Auto, Prone

3” Mortar: Mortar positions were quick to set up in battle, and provided a steady rain of fire on the enemy.

6 pdr Anti-Tank Gun: First used in Africa, the six pounder was a versatile anti-tank gun, though the armour of later German tanks would soon prove an effective defence from frontal attacks.

Anti-Tank Explosives: A combination of high-powered grenades and make-shift charges, these explosives give infantry a fighting chance against armoured vehicles, though their use is only advised in built-up areas.

Bren LMG: Based on a Czech design, the Bren gun was easily recognised by its long, curved magazine, necessary for the .303 cartridge used by the British Army.

Flamethrower: The flamethrower was very effective at clearing infantry out of cover, and achieved a fearsome reputation.

Heavy Artillery: At the call of platoon leaders, heavy artillery could literally blast infantry out of cover, and was effective against all but the heaviest of armoured vehicles.

Improved Anti-Tank Explosives: Dedicated satchel charges were used by elite units in order to immobilise and destroy enemy armour, though such attacks were extremely dangerous.

Lee-Enfield rifle: 50 years old by the end of World War II, the Lee-Enfield remained a sturdy infantry weapon.

Light Artillery: Using smaller and short-ranged guns, light artillery is primarily an anti-infantry weapon, forcing enemy troops into cover and disrupting their attacks as an assault is launched.

PIAT: Though a clumsy weapon to use, the PIAT had smokeless projectiles and no backblast, allowing it to be fired safely from cover, even in the presence of observant tank crews.

Scoped Lee-Enfield rifle: Modified from the standard Lee-Enfield, this was a weapon of choice for snipers.

Sten SMG: The Sten was designed to be quick, easy and cheap to manufacture, but remained potent on the battlefield.

Vickers MMG: The Vickers was simple, reliable and efficient. It was perhaps a perfect support weapon.

LATE WAR BRITISH ARMY PARATROOPERS

The Parachute Regiment (the Red Devils) has a stringent selection process and its soldiers are trained to operate behind enemy lines with little or no support. The most famous airborne operation the Parachute Regiment took part in was Operation Market Garden, an attempt to capture a number of bridges across the Rhine in Holland to allow free passage over the river into the heart of Germany. Though partially successful, the final crossing at Arnhem was not captured – the bridge too far.

Platoon Structure

The standard section size for most British Army platoons is ten soldiers, consisting of two teams. The section is led by a Corporal, while the second team is led by a Lance Corporal.

Three such sections make up a standard platoon, led by a Lieutenant, who is in turn assisted by a Sergeant. Three platoons make a rifle (infantry) company.

A platoon that has been on operational duty for some time may vary wildly from this, however. Heavy losses can result in fewer and smaller sections, while a platoon involved in heavy combat may become 'reinforced' with more soldiers, giving it more sections.

You can make any platoon with the British Army list, from one that has just been deployed, to a grizzled set of veterans that have taken heavy losses or been reinforced.

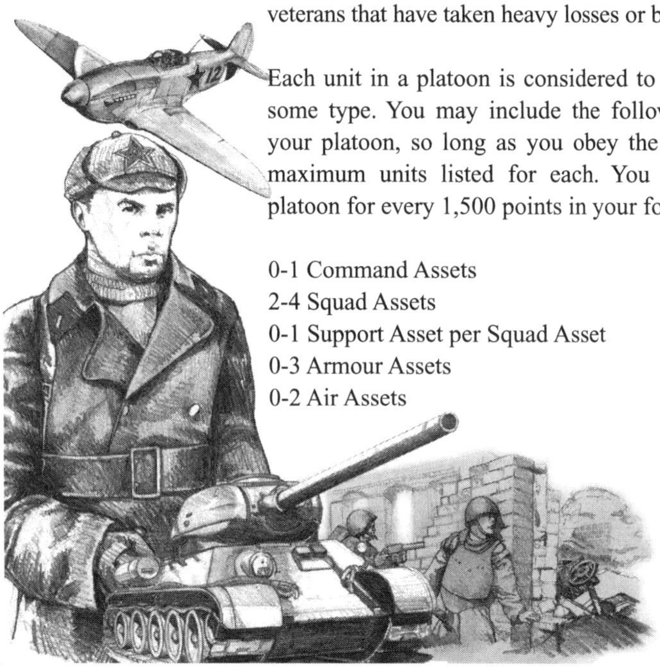

Each unit in a platoon is considered to be an Asset of some type. You may include the following Assets in your platoon, so long as you obey the minimum and maximum units listed for each. You may take one platoon for every 1,500 points in your force.

0-1 Command Assets
2-4 Squad Assets
0-1 Support Asset per Squad Asset
0-3 Armour Assets
0-2 Air Assets

For details on more Support, Armour and Air Assets see the British Vehicle and Air Assets List on page 61).

Special Rules

The following special rules apply to all British Paratrooper platoons.

Elite Training

All British Paratrooper infantry units gain a +1 bonus to their Armour Saves when in Cover. They may also re-roll their Close Combat dice.

Fearless

Every British Paratrooper infantry unit discards the first Suppression dice it suffers every turn.

Fire Discipline

British Paratrooper infantry units need only discard one Damage Dice for each extra Fire Zone created.

Parachute Force Deployment

A British Army Paratroopers platoon may always deploy by parachute (see page 38). If you choose to do so, all units in the platoon capable of deploying this way must do so – any units that cannot deploy this way may not be used. The platoon may also take up to three Air Assets rather than two.

Command Assets

The following are all the Command Assets available to a British Army Paratrooper force.

Paratrooper Command Section - 90 points

Leading his soldiers into battle, the Lieutenant takes the same risks as his men when dropping behind enemy lines.

Type	Size	Move	Close Combat	Target	Save	Kill	Traits
Lieutenant	1	5"	D6+1	4+	6+	6+	-
Sergeant	1	5"	D6+1	4+	6+	6+	-
Soldier	1	5"	D6+1	4+	6+	6+	-

Unit: Lieutenant (unit leader) with a Sten SMG, Sergeant with a Lee-Enfield rifle, 2 Soldiers with Lee-Enfield rifles.

Options: This unit may be deployed by parachute. The Sergeant may swap his rifle for a Sten SMG for +0 points. A Bren Team may be added to the Command Section for +75 points. An Anti-Tank Team may be added to the Command Section for +60 points. These extra teams may join the Command Section as one unit, or be split off to act as separate units.

Supporting Artillery: This unit may purchase up to 3 uses of Supporting Artillery. Calling in Supporting Artillery requires a Ready action. This is used as an Artillery attack, but the initial Impact Point must be in the unit's Line of Sight. There is no minimum range for this attack (yes, you can call it down right on your own head!). Light Artillery costs +50 points per use, while Heavy Artillery costs +100 points per use.

Squad Assets

The following are all the Squad Assets available to a British Army Paratroopers force.

Paratroopers Section - 230 points

Tasked with operating behind enemy lines, Paratroopers are taught self-reliance and personal initiative, the clear tools to survival on their missions.

Type	Size	Move	Close Combat	Target	Save	Kill	Traits
Corporal	1	5"	D6+1	4+	6+	6+	-
Lance Corporal	1	5"	D6+1	4+	6+	6+	-
Soldier	1	5"	D6+1	4+	6+	6+	-

Team 1: Corporal (unit leader) with a Sten SMG, 6 Soldiers with Lee-Enfield rifles.

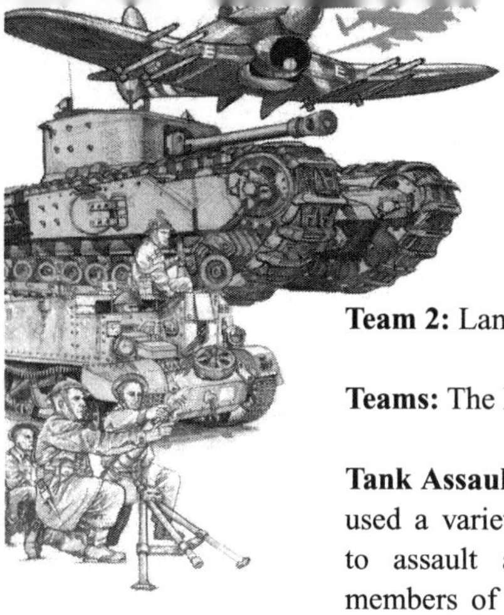

Team 2: Lance Corporal with a Sten SMG, Soldier with a Lee-Enfield rifle, Soldier with a Bren LMG.

Teams: The Lance Corporal may act as unit leader to Team 2, splitting off to act as a separate unit.

Tank Assault: Infantry in this period used a variety of grenades or mines to assault armoured vehicles. All members of the unit are assumed to carry Anti-Tank Explosives. These may be placed on any model of Size 2 or greater with a Charge action if the target ends up within the Lethal Range of the charging unit. They will explode at the end of the charging unit's next turn.

Options: This unit may be deployed by parachute. All Soldiers may swap their Lee-Enfield rifles for Sten SMGs for +0 points. One Soldier may swap his Lee-Enfield rifle for a Scoped Lee-Enfield rifle for +10 points.

Support Assets

The following are all the Support Assets available to a British Army Paratroopers force.

Paratroopers Anti-Tank Team - 60 points

The ability to include anti-tank soldiers in an airborne force means that even behind their front line, armoured crews can never be completely safe.

Type	Size	Move	Close Combat	Target	Save	Kill	Traits
Soldier	1	4"	D6+1	4+	6+	6+	-

Unit: Soldier (unit leader) with a PIAT, Soldier with a Lee-Enfield rifle.

Options: This unit may be deployed by parachute.

Paratroopers Bren Team - 75 points

Roving the battlefield, this team lends a hail of supporting fire where it is most needed, before redeploying to lend assistance to another attack.

Type	Size	Move	Close Combat	Target	Save	Kill	Traits
Soldier	1	5"	D6+1	4+	6+	6+	-

Unit: Soldier (unit leader) with a Bren, Soldier with a Lee-Enfield rifle.

Options: This unit may be deployed by parachute.

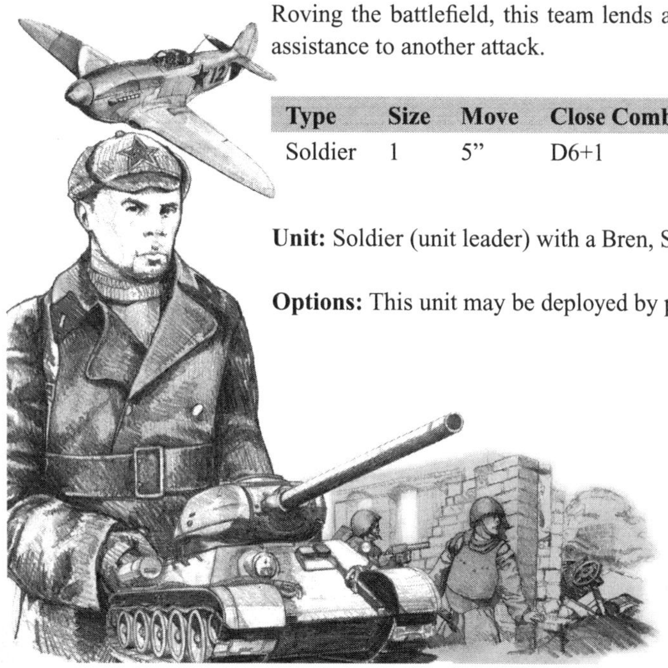

Paratroopers Sniper Team – 95 points

Deploying ahead of the main strike force, the sniper team chooses its location with care, targeting vital enemy personnel to cause chaos and confusion.

Type	Size	Move	Close Combat	Target	Save	Kill	Traits
Soldiers	1	5"	D6+1	4+	6+	6+	-

Unit: Soldier (unit leader) with a Scoped Lee-Enfield rifle, Soldier with a Scoped Lee-Enfield rifle.

Stealthy: If a model from this unit is in Cover and did not make a Shoot action or reaction in the last turn, enemy models more than 20" away may not draw Line of Sight to it.

Ambush: This unit may be concealed in an area of terrain outside the enemy's deployment zone before the game begins. Note the location of the unit before any models are deployed on the table. You can reveal the ambush by placing the section on the table at any time after a unit (friendly or otherwise) has completed an action. However, the unit must be revealed by the end of turn three or be counted as destroyed. You may not use the Ambush rules if your force is deploying by parachute.

Options: This unit may be deployed by parachute.

Paratroopers Medium Mortar Team – 95 points

This small three man team is capable of laying down a wave of short-ranged bombardment, forcing enemies into cover.

Type	Size	Move	Close Combat	Target	Save	Kill	Traits
Corporal	1	4"	D6+1	4+	6+	6+	-
Soldier	1	4"	D6+1	4+	6+	6+	-

Unit: Corporal (unit leader) with a Lee-Enfield rifle, Soldier with a 3" Mortar (FaF), Soldier with a Lee-Enfield rifle.

Options: This unit may be deployed by parachute. The Corporal may swap the rifle for a Sten SMG for +0 points.

British Paratroopers

Paratroopers MMG Support Team - 105 points

When a Bren team just will not do, the Vickers offers a heavier weight of fire, ripping apart the most determined assaults.

Type	Size	Move	Close Combat	Target	Save	Kill	Traits
Corporal	1	4”	D6+1	4+	6+	6+	-
Soldier	1	4”	D6+1	4+	6+	6+	-

Unit: Corporal (unit leader) with a Lee-Enfield rifle, Soldier with a Vickers MMG, Soldier with a Lee-Enfield rifle.

Options: This unit may be deployed by parachute.

Paratroopers 6 pdr Anti-Tank Gun - 105 points

Only deployed after the parachute drop, the six pounder lends a greater anti-tank capability to the Parachute Regiment.

Type	Size	Move	Close Combat	Target	Save	Kill	Traits
Sergeant	1	4”	D6+1	4+	6+	6+	-
Soldier	1	4”	D6+1	4+	6+	6+	-

Unit: Sergeant (unit leader) with a Sten SMG, Soldier with a 6 pdr Anti-Tank Gun (FaF), Soldier with a Lee-Enfield rifle.

Crew Weapon: The 6 pdr Anti-Tank Gun can only shoot if at least one crew member is in base contact with it.

Cumbersome: A Ready action must be taken before the 6 pdr Anti-Tank Gun may be moved. The Move score of all models in the unit depends on the number of models still manning the weapon.

3 crew = 3” maximum move,
2 crew = 2” maximum move and
1 crew = No movement possible.

If the crew choose to abandon the 6 pdr Anti-Tank Gun, they may move normally.

Gun Shield: The Sergeant and Soldiers count as being in Medium Cover when shot from the Forward facing.

Options: This unit may have a Bedford Truck for +50 points or an M3 halftrack for +80 points (see US vehicle and Air Assets list).

Engineer Section – 135 points

The Engineers of the Parachute Regiment are highly capable soldiers, able to lend their skills to the most critical focus on the battlefield.

Type	Size	Move	Close Combat	Target	Save	Kill	Traits
Sergeant	1	5”	D6+1	4+	6+/6+	6+	-
Corporal	1	5”	D6+1	4+	6+	6+	-
Soldier	1	5”	D6+1	4+	6+	6+	-

Team 1: Sergeant (unit leader) with a Lee-Enfield rifle, 2 Soldiers with Lee-Enfield rifles.

Team 2: Corporal with a Flamethrower, 2 Soldiers with Lee-Enfield rifles.

Teams: The Corporal may act as unit leader to Team 2, splitting off to act as a separate unit.

Ambush: This unit may be concealed in an area of terrain outside the enemy's deployment zone before the game begins. Note the location of the unit before any models are deployed on the table. You can reveal the ambush by placing the section on the table at any time after a unit (friendly or otherwise) has completed an action. However, the unit must be revealed by the end of turn three or be counted as destroyed. You may not use the Ambush rules if your force is deploying by parachute or with a transport vehicle.

Tank Assault: Infantry in this period used a variety of grenades or mines to assault armoured vehicles. All members of the unit are assumed to carry Anti-Tank Explosives. These may be placed on any model of Size 2 or greater with a Charge action if the target ends up within the Lethal Range of the charging unit. They will explode at the end of the charging unit's next turn.

Options: Any member of the unit may swap their Lee-Enfield rifle for a Sten SMG for +0 points. The whole unit may be equipped with Improved Anti-Tank Explosives for +30 points. An Engineer Section may have 2 Bren Carriers for +140 points, or a Bedford Truck for +50 points, if it picks either it may not use the Ambush rule.

Late War British Armoury

Weapon	Range	Damage	Traits
3“ Mortar	48”	D6	Artillery, Lethal Zone/3”, Ready, Slow
6 pdr Anti-Tank Gun	36”	D10+2	Lethal Zone/1“, Ready, Slow
Anti-Tank Explosives	Placed	D10	Lethal Zone/2“, One-Shot, Piercing/2
Bren LMG	24”	2xD6	Auto, Prone
Flamethrower	10”	3xD6	Flame, Retaliate, Slow
Heavy Artillery	Special	2xD6+1	Lethal Zone/4“, One-Shot
Improved Anti-Tank Explosives	Placed	2xD10+2	Lethal Zone/2“, One-Shot, Piercing/3
Lee-Enfield rifle	24”	D6	-
Light Artillery	Special	D6	Lethal Zone/4“, One-Shot
PIAT	18”	D6+3	Lethal Zone/ 1“, Piercing/1, Slow
Scoped Lee-Enfield rifle	36”	D6	Accurate
Sten SMG	12”	3xD6	-
Vickers MMG	48”	2xD6+1	Auto, Prone

3” Mortar: Mortar positions were quick to set up in battle, and provided a steady rain of fire on the enemy.

6 pdr Anti-Tank Gun: First used in Africa, the six pounder was a versatile anti-tank gun, though the armour of later German tanks would soon prove an effective defence from frontal attacks.

Anti-Tank Explosives: A combination of high-powered grenades and make-shift charges, these explosives give infantry a fighting chance against armoured vehicles, though their use is only advised in built-up areas.

Bren LMG: Based on a Czech design, the Bren gun was easily recognised by its long, curved magazine, necessary for the .303 cartridge used by the British Army.

Flamethrower: The flamethrower was very effective at clearing infantry out of cover, and achieved a fearsome reputation.

Heavy Artillery: At the call of platoon leaders, heavy artillery could literally blast infantry out of cover, and was effective against all but the heaviest of armoured vehicles.

Improved Anti-Tank Explosives: Dedicated satchel charges were used by elite units in order to immobilise and destroy enemy armour, though such attacks were extremely dangerous.

Lee-Enfield rifle: 50 years old by the end of World War II, the Lee-Enfield remained a sturdy infantry weapon.

Light Artillery: Using smaller and short-ranged guns, light artillery is primarily an anti-infantry weapon, forcing enemy troops into cover and disrupting their attacks as an assault is launched.

PIAT: Though a clumsy weapon to use, the PIAT had smokeless projectiles and no backblast, allowing it to be fired safely from cover, even in the presence of observant tank crews.

Scoped Lee-Enfield rifle: Modified from the standard Lee-Enfield, this was a weapon of choice for snipers.

Sten SMG: The Sten was designed to be quick, easy and cheap to manufacture, but remained potent on the battlefield.

Vickers MMG: The Vickers was simple, reliable and efficient. It was perhaps a perfect support weapon.

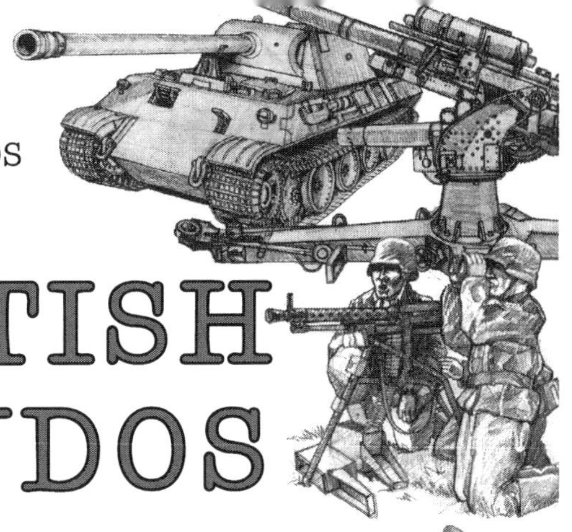

LATE WAR BRITISH ARMY COMMANDOS

Initially created as a raiding force, dedicated to disrupting enemy logistics and communications, the Commandos soon achieved legendary status and began to take the role of shock troops. Perhaps their most famous mission was the attack on St. Nazaire in Operation Chariot, but they were also active in France, Germany, Norway and Burma.

Platoon Structure

The standard section size for most British Army Commando platoons is nine soldiers, consisting of two teams. The section is led by a Corporal, while the second team is led by a Lance Corporal.

Three such sections make up a standard platoon, led by a Lieutenant, who is in turn assisted by a Sergeant. Three platoons make a rifle (infantry) company.

A platoon that has been on operational duty for some time may vary wildly from this, however. Heavy losses can result in fewer and smaller sections, while a platoon involved in heavy combat may become 'reinforced' with more soldiers, giving it more sections.

You can make any platoon with the British Army list, from one that has just been deployed, to a grizzled set of veterans that have taken heavy losses or been reinforced.

Each unit in a platoon is considered to be an Asset of some type. You may include the following Assets in your platoon, so long as you obey the minimum and maximum units listed for each. You may take one platoon for every 1,500 points in your force.

0-1 Command Assets
2-4 Squad Assets
0-1 Support Asset per Squad Asset
0-3 Armour Assets
0-2 Air Assets

For details on more Support, Armour and Air Assets see the British Vehicle and Air Assets List on page 61).

Special Rules

The following special rules apply to all British Commando platoons.

Elite Training

All British Commando infantry units gain a +1 bonus to their Armour Saves when in Cover. They may also re-roll their Close Combat dice.

Fearless

Every British Commando infantry unit discards the first Suppression dice it suffers every turn.

Fire Discipline

British Commando infantry units need only discard one Damage Dice for each extra Fire Zone created.

Stealthy

If a model from a British Commando infantry unit is in Cover and did not make a Shoot action or reaction in the last turn, enemy models more than 20" away may not draw Line of Sight to it.

Ambush

Any one British Commando infantry unit may be concealed in an area of terrain outside the enemy's deployment zone before the game begins. Note the location of the unit before any models are deployed on the table. You can reveal the ambush by placing the section on the table at any time after a unit (friendly or otherwise) has completed an action. However, the unit must be revealed by the end of turn three or be counted as destroyed.

Command Assets

The following are all the Command Assets available to a British Army Commando force.

Commando Command Section - 105 points

The Command section leads the Platoon into battle.

Type	Size	Move	Close Combat	Target	Save	Kill	Traits
Lieutenant	1	4”	D6+2	4+	6+	6+	-
Sergeant	1	4”	D6+2	4+	6+	6+	-
Soldier	1	4”	D6+2	4+	6+	6+	-

Unit: Lieutenant (unit leader) with a Sten SMG, Sergeant with a Lee-Enfield rifle, 2 Soldiers with Lee-Enfield rifles.

Options: The Sergeant may swap his rifle for a SMG for + 0 points. A Bren Team may be added to the Command Section for +85 points. An Anti-Tank Team may be added to the Command Section for +70 points. These extra teams may join the Command Section as one unit, or be split off to act as separate units.

Supporting Artillery: This unit may purchase up to 3 uses of Supporting Artillery. Calling in Supporting Artillery requires a Ready action. This is used as an Artillery attack, but the initial Impact Point must be in the unit’s Line of Sight. There is no minimum range for this attack (yes, you can call it down right on your own head!). Light Artillery costs +50 points per use, while Heavy Artillery costs +100 points per use.

Squad Assets

The following are all the Squad Assets available to a British Army Commando force.

Commando Squad - 250 points

The Commandos recruited and trained the best of the best, and were able to succeed in the most perilous of missions.

Type	Size	Move	Close Combat	Target	Save	Kill	Traits
Corporal	1	4”	D6+2	4+	6+	6+	-
Lance Corporal	1	4”	D6+2	4+	6+	6+	-
Soldier	1	4”	D6+2	4+	6+	6+	-

Team 1: Corporal (unit leader) with a Sten SMG, 6 Soldiers with Lee-Enfield rifles.

Team 2: Lance Corporal with a Sten SMG, Soldier with a Lee-Enfield rifle, Soldier with a Bren LMG.

Teams: The Lance Corporal may act as unit leader to Team 2, splitting off to act as a separate unit.

Tank Assault: Infantry in this period used a variety of grenades or mines to assault armoured vehicles. All members of the unit are assumed to carry Anti-Tank Explosives. These may be placed on any model of Size 2 or greater with a Charge action if the target ends up within the Lethal Range of the charging unit. They will explode at the end of the charging unit's next turn.

Options: All Soldiers may swap the Lee-Enfield rifle for a Sten SMG for +0 points.

Support Assets

The following are all the Support Assets available to a British Army Commando force.

Commando Anti-Tank Team - 70 points

The combination of a Commando and a PIAT is a fearsome one, and tank crews are advised to be wary of sudden ambushes to the side and rear.

Type	Size	Move	Close Combat	Target	Save	Kill	Traits
Soldier	1	4"	D6+2	4+	6+	6+	-

Unit: Soldier (unit leader) with a PIAT, Soldier with a Lee-Enfield rifle.

Commando Bren Team - 85 points

Taking position on the flanks of the main force, Bren teams lay down a hail of fire to support the advance of their allies.

Type	Size	Move	Close Combat	Target	Save	Kill	Traits
Soldier	1	4"	D6+2	4+	6+	6+	-

Unit: Soldier (unit leader) with a Bren, a Soldier with a Lee-Enfield rifle.

Commando Sniper Team - 80 points

There are few things more dangerous on the battlefield than a Commando with a scoped Lee-Enfield. Enemy officers and machine gunners are the prime targets.

Type	Size	Move	Close Combat	Target	Save	Kill	Traits
Soldier	1	5"	D6+2	4+	6+	6+	-

Unit: Soldier (unit leader) with a Scoped Lee-Enfield rifle, a Soldier with Scoped Lee-Enfield rifle.

British Commandos

Commando Medium Mortar Team - 110 points

Commandos need battlefield support as much as the next force and mobile mortars provide a suitable means of launching light strikes against an enemy.

Type	Size	Move	Close Combat	Target	Save	Kill	Traits
Corporal	1	4”	D6+2	4+	6+	6+	-
Soldier	1	4”	D6+2	4+	6+	6+	-

Unit: Corporal (unit leader) with a Lee-Enfield rifle, Soldier with a 3” Mortar (FaF), Soldier with a Lee-Enfield rifle.

Options: The Corporal may swap his Lee-Enfield rifle for a Sten SMG for +0 points.

Commando MMG Support Team - 120 points

Armed with a Vickers medium machine gun, this team is able to lay down a terrible hail of fire that will drive back an attack and force enemy infantry to keep their heads down.

Type	Size	Move	Close Combat	Target	Save	Kill	Traits
Corporal	1	4”	D6+2	4+	6+	6+	-
Soldier	1	4”	D6+2	4+	6+	6+	-

Unit: Corporal (unit leader) with a Lee-Enfield rifle, Soldier with a Vickers MMG, Soldier with a Lee-Enfield rifle.

Commando 6 pdr Anti-Tank Gun - 125 points

Providing a vital long-ranged anti-tank capability, the presence of a 6 pounder ensures the Commandos have little to fear on the battlefield.

Type	Size	Move	Close Combat	Target	Save	Kill	Traits
Sergeant	1	4”	D6+1	4+	6+	6+	-
Soldier	1	4”	D6+1	4+	6+	6+	-

Unit: Sergeant (unit leader) with Sten SMG, Soldier with AT Gun (FaF), Soldier with Lee-Enfield rifle.

Crew Weapon: The 6 pdr Anti-Tank Gun can only shoot if at least one crew member is in base contact with it.

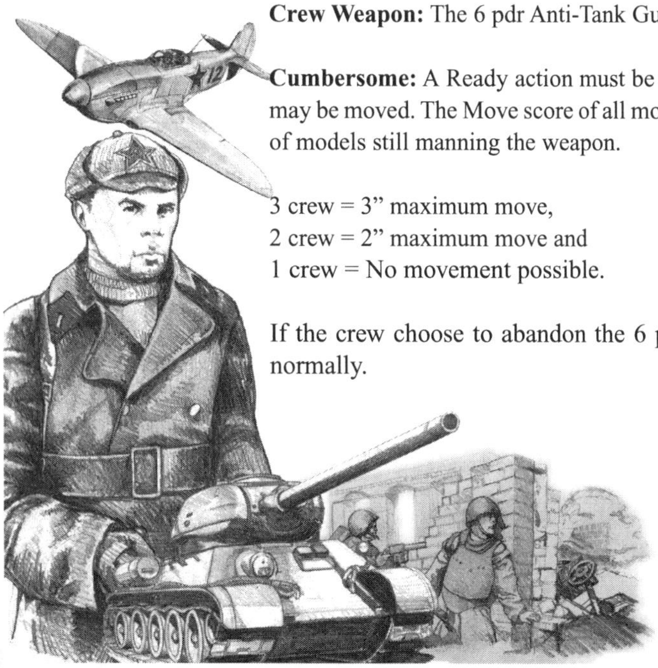

Cumbersome: A Ready action must be taken before the 6 pdr Anti-Tank Gun may be moved. The Move score of all models in the unit depends on the number of models still manning the weapon.

3 crew = 3” maximum move,
2 crew = 2” maximum move and
1 crew = No movement possible.

If the crew choose to abandon the 6 pdr Anti-Tank Gun, they may move normally.

Gun Shield: The Sergeant and Soldiers count as being in Medium Cover when shot from the Forward facing.

Options: This unit may have a Bedford Truck for +50 points or an M3 halftrack for +80 points (see US vehicle and Air Assets list).

Engineer Section - 150 points

Roaming the battlefield, the Engineers provide ad hoc support in close assaults or against armoured vehicles.

Type	Size	Move	Close Combat	Target	Save	Kill	Traits
Sergeant	1	5"	D6+1	4+	6+	6+	-
Corporal	1	5"	D6+1	4+	6+	6+	-
Soldier	1	5"	D6+1	4+	6+	6+	-

Team 1: Sergeant (unit leader) with a Lee-Enfield rifle, 2 Soldiers with Lee-Enfield rifles.

Team 2: Corporal with a Flamethrower, 2 Soldiers with Lee-Enfield rifles.

Teams: The Corporal may act as unit leader to Team 2, splitting off to act as a separate unit.

Tank Assault: Infantry in this period used a variety of grenades or mines to assault armoured vehicles. All members of the unit are assumed to carry Anti-Tank Explosives. These may be placed on any model of Size 2 or greater with a Charge action if the target ends up within the Lethal Range of the charging unit. They will explode at the end of the charging unit's next turn.

Options: Any member of the unit may swap their Lee-Enfield rifle for a Sten SMG for +0 points. The whole unit may be equipped with Improved Anti-Tank Explosives for +30 points. An Engineer Section may have 2 Universal Carriers for +140 points, or a Bedford Truck for +50 points. If it picks either it may not use the Ambush rule.

Late War British Armoury

Weapon	Range	Damage	Traits
3" Mortar	48"	D6	Artillery, Lethal Zone/3", Ready, Slow
6 pdr Anti-Tank Gun	36"	D10+2	Lethal Zone/1", Ready, Slow
Anti-Tank Explosives	Placed	D10	Lethal Zone/2", One-Shot, Piercing/2
Bren LMG	24"	2xD6	Auto, Prone
Flamethrower	10"	3xD6	Flame, Slow
Heavy Artillery	Special	2xD6+1	Lethal Zone/4", One-Shot
Improved Anti-Tank Explosives	Placed	2xD10+2	Lethal Zone/2", One-Shot, Piercing/3
Lee-Enfield Rifle	24"	D6	-
Light Artillery	Special	D6	Lethal Zone/4", One-Shot
PIAT	18"	D6+3	Lethal Zone/1", Piercing/1
Scoped Lee-Enfield	36"	D6	Accurate
Sten SMG	12"	3xD6	-
Vickers MMG	48"	2xD6+1	Auto, Prone

3" Mortar: Mortar positions were quick to set up in battle, and provided a steady rain of fire on the enemy.

6 pdr Anti-Tank Gun: First used in Africa, the six pounder was a versatile anti-tank gun, though the armour of later German tanks would soon prove an effective defence from frontal attacks.

Anti-Tank Explosives: A combination of high-powered grenades and make-shift charges, these explosives give infantry a fighting chance against armoured vehicles, though their use is only advised in built-up areas.

Bren LMG: Based on a Czech design, the Bren gun was easily recognised by its long, curved magazine, necessary for the .303 cartridge used by the British Army.

Flamethrower: The flamethrower was very effective at clearing infantry out of cover, and achieved a fearsome reputation.

Heavy Artillery: At the call of platoon leaders, heavy artillery could literally blast infantry out of cover, and was effective against all but the heaviest of armoured vehicles.

Improved Anti-Tank Explosives: Dedicated satchel charges were used by elite units in order to immobilise and destroy enemy armour, though such attacks were extremely dangerous.

Lee-Enfield rifle: 50 years old by the end of World War II, the Lee-Enfield remained a sturdy infantry weapon.

Light Artillery: Using smaller and short-ranged guns, light artillery is primarily an anti-infantry weapon, forcing enemy troops into cover and disrupting their attacks as an assault is launched.

PIAT: Though a clumsy weapon to use, the PIAT had smokeless projectiles and no backblast, allowing it to be fired safely from cover, even in the presence of observant tank crews.

Scoped Lee-Enfield rifle: Modified from the standard Lee-Enfield, this was a weapon of choice for snipers.

Sten SMG: The Sten was designed to be quick, easy and cheap to manufacture, but remained potent on the battlefield.

Vickers MMG: The Vickers was simple, reliable and efficient. It was perhaps a perfect support weapon.

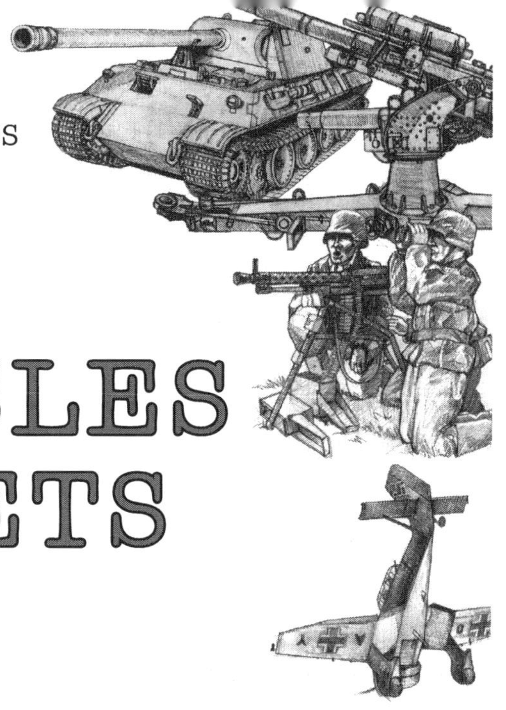

LATE WAR BRITISH VEHICLES AND AIR ASSETS

The following Assets may be used by any Late War British Army force.

Support Assets

The following are all the Support Assets available to a British Army force.

AEC Mk III – 160 points

An armoured car based on the Matador chassis that towed field guns, the Mk III mounted a 75mm tank gun.

Type	Size	Move	Close Combat	Facing	Target	Save	Kill	Traits
AEC Mk III	2	6"	2xD10	Front	8+	4+	10+	Hits/2, Lumbering, Multifire/2, Wheels/11"
				Sides	7+	5+	9+	
				Rear	7+	5+	8+	

Unit: AEC Mk III with a 75mm tank gun and an MG.

Co-Axial MG: To improve target acquisition, MGs on vehicles were often mounted co-axial to the main gun. The co-axial MG must always use the same Fire Zone as the main gun. As soon as a Fire Zone is placed within range of the co-axial MG, you may add +1 to all Damage Dice of the main gun.

Bedford Truck – 50 points

Based on a 1.5 tonne chassis, the Bedford Truck was the British Army's standard transport vehicle, moving supplies, weapons and men across Europe.

Type	Size	Move	Close Combat	Facing	Target	Save	Kill	Traits
Medium Truck	4	5"	D10	Front/ Sides/ Rear	6+	6+	7+	Hits/3, Lumbering, Wheels/10"

Unit: Bedford Truck.

Transport: The Bedford Truck can carry up to 10 Size points of models (each maximum Size 1).

British Vehicles

Bofors 40mm AA Portee - 100 points

One of the most widely-used medium anti-aircraft systems during World War II, the Bofors was capable of bringing down any fighter-bomber that dared to enter the surrounding area.

Type	Size	Move	Close Combat	Facing	Target	Save	Kill	Traits
Bofors Portee	4	5"	D10	Front/ Sides/ Rear	6+	6+	8+	Hits/2, Lumbering, Wheels/10"

Unit: Portee with a 40mm Bofors AA cannon (FaF).

Universal Carrier - 70 points

Correctly known as the Universal Carrier, but typically called the Bren Carrier even if no Bren gun was installed, this armoured vehicle was widely used by Allied forces. Developed to transport both troops and support weapons, it was to become the most numerous armoured vehicle ever built.

Type	Size	Move	Close Combat	Facing	Target	Save	Kill	Traits
Universal Carrier	2	5"	2xD10	Front	7+	5+	9+	Hits/2, Lumbering, Tracks/10"
				Sides	6+	6+	8+	
				Rear	6+	6+	7+	

Unit: Universal Carrier with a Pintle MG (FaF).

Transport: The Universal Carrier can carry up to 5 Size points of models (each maximum Size 1).

Daimler Dingo MkII scout car - 95 points

A light four wheeled-drive vehicle, the Dingo was used in the reconnaissance role, and was surprisingly well-armoured.

Type	Size	Move	Close Combat	Facing	Target	Save	Kill	Traits
Dingo Scout Car	2	9"	2xD10	Front	7+	4+	9+	Hits/2, Lumbering, Wheels/18"
				Sides	7+	5+	9+	
				Rear	6+	6+	7+	

Unit: Dingo Scout Car with a Pintle MG (FaF).

Humber Scout Car- 90 points

Similar to the Dingo, the Humber was created because Daimler could not build enough of the former. Though lacking a little in speed, it was comparable to the Dingo in most respects.

Type	Size	Move	Close Combat	Facing	Target	Save	Kill	Traits
Humber Scout Car	2	8"	2xD10	Front	7+	4+	9+	Hits/2, Lumbering, Wheels/15"
				Sides	7+	5+	9+	
				Rear	6+	6+	7+	

Unit: Humber Scout Car with a Pintle MG (FaF).

M3 Halftrack - 80 points

An American-built armoured vehicle, the M3 was used in great numbers by the British Army, and had dozens of different variants.

Type	Size	Move	Close Combat	Facing	Target	Save	Kill	Traits
M3 Halftrack	3	7"	2xD10	Front	7+	5+	8+	Hits/2, Lumbering, Tracks/14"
				Sides	6+	6+	7+	
				Rear	6+	6+	7+	

Unit: M3 with a Pintle MG (FaF).

Transport: The M3 can carry up to 12 Size points of models (each maximum Size 1).

Options: A Pintle MG (FaA) may be added for +35 points.

Wasp Flamethrower - 70 points

A variant of the Bren Carrier, the Wasp was a lethal vehicle, able to flush infantry out of cover with a jet of burning flame.

Type	Size	Move	Close Combat	Facing	Target	Save	Kill	Traits
Wasp	2	5"	2xD10	Front	7+	5+	9+	Hits/2, Lumbering, Retaliate, Tracks/10"
				Sides	6+	6+	8+	
				Rear	6+	6+	7+	

Unit: Wasp Carrier with a Heavy Flamethrower (FaF).

Armour Assets

The following are all the Armour Assets available to a British Army force.

Achilles Tank Destroyer – 225 points

Based on the American M10 Wolverine, this British variant mounted the highly successful 17-pounder anti-tank gun in a modified turret.

Type	Size	Move	Close Combat	Facing	Target	Save	Kill	Traits
Achilles	4	4"	4xD10	Front	7+	4+	11+	Hits/2, Lumbering, Tracks/8"
				Sides	7+	5+	8+	
				Rear	7+	5+	7+	

Unit: Achilles with a 17 pdr gun and a Browning .50 cal MG.

Browning .50 cal MG: This weapon may not be fired in the same Shoot action as the main gun.

Churchill Crocodile Infantry Tank – 330 points

Retaining the Churchill's main gun, the Crocodile replaced one machine gun with a flamethrower, making it both resilient and lethal to enemy infantry.

Type	Size	Move	Close Combat	Facing	Target	Save	Kill	Traits
Churchill Crocodile	4	3"	4xD10	Front	9+	3+	12+	Hits/3, Lumbering, Multifire/3, Retaliate, Tracks/5"
				Sides	8+	4+	11+	
				Rear	7+	4+	10+	

Unit: Churchill Crocodile with a 75mm tank gun, an MG, a Pintle MG and a Heavy Flamethrower (FaF).

Co-Axial MG: To improve target acquisition, MGs on vehicles were often mounted co-axial to the main gun. The co-axial MG must always use the same Fire Zone as the main gun. As soon as a Fire Zone is placed within range of the co-axial MG, you may add +1 to all Damage Dice of the main gun.

British Vehicles

Churchill VII Infantry Tank - 330 points

Very heavily armoured, this version of the Churchill eliminated many of the reliability problems that plagued its predecessors.

Type	Size	Move	Close Combat	Facing	Target	Save	Kill	Traits
Churchill VII	4	3”	4xD10	Front	9+	3+	12+	Hits/3, Lumbering, Multifire/3, Tracks/5”
				Sides	8+	4+	11+	
				Rear	7+	4+	10+	

Unit: Churchill VII with a 75mm tank gun, an MG, an MG (FaF), and a Pintle MG.

Co-Axial MG: To improve target acquisition, MGs on vehicles were often mounted co-axial to the main gun. One MG has a co-axial mount. The co-axial MG must always use the same Fire Zone as the main gun. As soon as a Fire Zone is placed within range of the co-axial MG, you may add +1 to all Damage Dice of the main gun.

Cromwell VII Cruiser Tank - 170 points

By the late period of the war, this was the most numerous British tank, and it was a highly successful design.

Type	Size	Move	Close Combat	Facing	Target	Save	Kill	Traits
Cromwell VII	4	5”	4xD10	Front	9+	4+	12+	Hits/2, Lumbering, Multifire/2, Tracks/10”
				Sides	8+	5+	11+	
				Rear	7+	5+	10+	

Unit: Cromwell VII with a 75mm tank gun and an MG.

Co-Axial MG: To improve target acquisition, MGs on vehicles were often mounted co-axial to the main gun. The co-axial MG must always use the same Fire Zone as the main gun. As soon as a Fire Zone is placed within range of the co-axial MG, you may add +1 to all Damage Dice of the main gun.

British Vehicles

Sherman Firefly VC Medium Tank - 285 points

A British version of the American Sherman, the Firefly mounted a 17 pounder main gun, and was one of the few Allied tanks capable of engaging a Tiger on an almost even basis.

Type	Size	Move	Close Combat	Facing	Target	Save	Kill	Traits
Sherman Firefly	4	4"	4xD10	Front	8+	4+	12+	Hits/2, Lumbering, Multifire/2, Tracks/8"
				Sides	7+	5+	11+	
				Rear	7+	5+	10+	

Unit: Sherman Firefly with a 17 pdr gun, a Browning .50 cal MG, an MG, and an MG (FaF).

Browning .50 cal MG: This weapon may not be fired in the same Shoot action as the main gun. When fired it may create its own separate Fire Zone without discarding any Damage Dice.

Co-Axial MG: To improve target acquisition, MGs on vehicles were often mounted co-axial to the main gun. One MG has a co-axial mount. The co-axial MG must always use the same Fire Zone as the main gun. As soon as a Fire Zone is placed within range of the co-axial MG, you may add +1 to all Damage Dice of the main gun.

Sherman III Medium Tank - 210 points

With many different variants, the Sherman was America's main medium tank in the war, and this variant was common in British forces.

Type	Size	Move	Close Combat	Facing	Target	Save	Kill	Traits
Sherman III	4	3"	4xD10	Front	8+	4+	12+	Hits/2, Lumbering, Multifire/2, Tracks/7"
				Sides	7+	5+	11+	
				Rear	7+	5+	10+	

Unit: Sherman III with a 75mm tank gun, an MG (FaF), an MG, and a Pintle MG.

Co-Axial MG: To improve target acquisition, MGs on vehicles were often mounted co-axial to the main gun. One MG has a co-axial mount. The co-axial MG must always use the same Fire Zone as the main gun. As soon as a Fire Zone is placed within range of the co-axial MG, you may add +1 to all Damage Dice of the main gun.

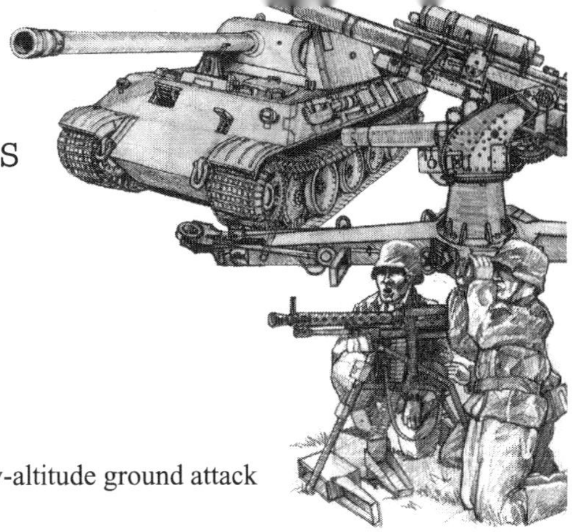

Air Assets

The following are all the Air Assets available to a British Army force.

Hawker Typhoon IB - 260 points

Initially designed as a high-altitude interceptor, the Typhoon would become one of the finest low-altitude ground attack aircraft of the war.

Type	Size	Move	Close Combat	Target	Save	Kill	Traits
Typhoon IB	5	Loiter Cruise	4xD10	6+	5+/5+	8+	Hits/3, Multifire/4

Unit: Hawker Typhoon IB with four Hispano 20mm cannon (FaF) and two 500 lb. bombs.

500 lb. Bomb: The Typhoon carries two 500 lb. bombs. They are dropped weapons, as detailed on page 16. Up to two may be dropped with every Shoot (bomb) action.

British Vehicle Armoury

Weapon	Range	Damage	Traits
17 pdr gun	60”	D10+4	Lethal Zone/1”, Multihit, Piercing/1, Slow
40mm Bofors AA cannon	36”	4xD6+3	AA, Auto, Piercing/1, Slow
500 lb. Bomb	Dropped	D10	Lethal Zone/6”, Multihit, One Shot
75mm tank gun	48”	D10+2	Lethal Zone/1”, Slow
Browning .50 cal MG	30”	3xD6	AA, Auto, Piercing/1
Heavy Flamethrower	12”	3xD6+1	Flame, Slow
Hispano 20mm cannon	24”	D6+2	Piercing/1
MG	24”	3xD6	Auto
Pintle MG	24”	3xD6	AA, Auto

17 pdr gun: The best Allied anti-tank gun of the war, the 17 pounder could defeat any German armour at range.

40mm Bofors AA cannon: Based on a Swedish design the Bofors was a widespread aerial defence among Allied forces.

500 lb. Bomb: The standard medium-bomb used by fighter-bombers, accurate delivery could destroy armoured vehicles and tear apart bunkers.

75mm tank gun: Useful to support infantry, by the late war period, the 75mm tank gun was beginning to lose legs, with its shells too easily defeated by new and improved German armour.

Browning .50 cal MG: The Browning can be found on vehicles and aircraft throughout the 20th Century, and was very common as the main armament on Allied aircraft.

Heavy Flamethrower: If mounted in a vehicle, the fuel reserves and effective power of a flamethrower is greatly increased.

Hispano 20mm cannon: This 20mm cannon was a superb weapon that went on to form the armament of many aircraft.

MG: Covering a wide variety of weapons, machine guns are used to suppress enemy infantry and provide supporting fire.

Pintle MG: A pintel mounted machine gun can engage low-flying aircraft due to its greater field of fire and elevation.

LATE WAR GERMAN HEER GRENADIERS AND PANZERGRENADIERS

The Heer infantry force highlights the basic Grenadier as its greatest strength, though there plenty of supporting units. Though not as dreaded as the SS or respected as the Fallschirmjäger, the common German Grenadier still bore the most of the fighting in the closing days of the war. The addition of halftracks and other transports turned the common Grenadier into the renowned Panzergrenadier.

A Heer infantry force comprises one or more Kampfgruppen. The Kampfgruppe (plural Kampfgruppen) was an ad-hoc combined arms formation, usually employing combination of tanks, infantry, anti-tank weapons and artillery components, generally organised for a particular task or operation. A Kampfgruppe could range in size from a corps to a company, but most common was an Abteilung (battalion)-sized formation. Kampfgruppen were generally referred to by either their commanding officer or the parent division.

Platoon Structure

The standard squad size for most Grenadier platoons is 10 soldiers, consisting of two teams of five men each. The section is led by an Unteroffizier, while the second fire team is led by a Obergefreiter.

Three such sections make up a standard Grenadier Zug (platoon), led by a Leutnant, who is in turn assisted by a Unterfeldwebel.

A platoon that has been on operational duty for some time may vary wildly from this, however. Heavy losses can result in fewer and smaller sections, while a platoon involved in heavy combat may become 'reinforced' with more soldiers, giving it more sections.

You can make any platoon with the late war Grenadier army list, from one that has just been deployed, to a grizzled set of veterans that have taken heavy losses or been reinforced.

Each unit in a platoon is considered to be an Asset of some type. You may include the following Assets in your platoon, so long as you obey the minimum and maximum units listed for each. You may take one platoon for every 1,500 points in your force.

0-1 Command Assets
2-4 Squad Assets
0-3 Support Assets
0-3 Armour Assets
0-1 Air Assets

For details on more Support, Armour and Air Assets see the German Vehicle and Air Assets List on page 91).

Historical Note

The transport options for the Support Assets in this army list were only used by Panzergrenadier Kampfgruppen. For simplicity and clarity we have combined Grenadiers and Panzergrenadiers into one army list. It is up to you to decide how historically accurate you would like your force to be!

Command Assets

The following are all the Command Assets available to a Heer Grenadier force.

Grenadier Command Squad- 100 points

Both well-armed and versatile, the command squads of the Grenadiers were valuable front line units, as well as vital for the command structure of the army.

Type	Size	Move	Close Combat	Target	Save	Kill	Traits
Leutnant	1	4"	D6+1	4+	6+	6+	-
Unterfeldwebel	1	4"	D6+1	4+	6+	6+	-
Soldier	1	4"	D6+1	4+	6+	6+	-

Team 1: Leutnant (unit leader) with a MP-40 SMG, Unterfeldwebel with a MP-40 SMG, Soldier with a Kar-98k rifle.

Team 2: Soldier with a Panzerschreck, Soldier with a Kar-98k rifle.

Teams: One Soldier may act as unit leader to Fire Team 2, splitting off to act as a separate unit.

Options: Drop Team 2 for –65 points. Add up to one additional Team 2 for +65 points. A Grenadier Command Squad may have an Opel Blitz Medium Truck for +50 points, up to 2 Kuebelwagen for +35 points each, an SdKfz 251/10 halftrack for +110 points, or up to 2 SdKfz 250/1 Neu halftracks for +70 points each.

Supporting Artillery: This unit may purchase up to 1 use of Supporting Artillery. Calling in Supporting Artillery requires a Ready action. This is used as an Artillery attack, but the initial Impact Point must be in the unit's Line of Sight. There is no minimum range for this attack (yes, you can call it down right on your own head!). Light Artillery costs +50 points per use, while Heavy Artillery costs +100 points per use.

Squad Assets

The following are all the Squad Assets available to a Wehrmacht Grenadier force.

Grenadier Squad - 160 points

Well trained to attack using speed, German Grenadiers were often found where the fighting was at its fiercest.

Type	Size	Move	Close Combat	Target	Save	Kill	Traits
Unteroffizier	1	4"	D6+1	4+	6+	6+	-
Obergefreiter	1	4"	D6+1	4+	6+	6+	-
Soldier	1	4"	D6+1	4+	6+	6+	-

German Grenadiers

Team 1: Unteroffizier (unit leader) with a MP-40 SMG, 3 Soldiers with Kar-98k rifles, Soldier with an MG-42 LMG.

Team 2: Obergefreiter with a MP-40 SMG, 4 Soldiers with Kar-98k rifles.

Teams: The Obergefreiter may act as unit leader to Team 2, splitting off to act as a separate unit.

Tank Assault: Infantry in this period used a variety of grenades or mines to assault armoured vehicles. All members of the unit are assumed to carry Anti-Tank Explosives. These may be placed on any model of Size 2 or greater with a Charge action if the target ends up within the Lethal Range of the charging unit. They will explode at the end of the charging unit's next turn.

Options: Up to 1 Soldier of Team 2 may be equipped with an MG-42 LMG for +40 points. Up to 1 Soldier of either Team 1 or 2 may be equipped with a Panzerfaust in addition to his rifle for +30 points. Drop Team 2 for – 60 points. A Grenadier Squad may have an Opel Blitz Medium Truck for +50 points, an SdKfz 251/1 halftrack for +80 points, or up to 2 SdKfz 250/1 Neu halftracks for +70 points each.

Volksgrenadier Squad – 80 points

A last ditch effort of the German Army, the Volksgrenadiere were poorly trained. What they made up for in National Socialist ardour was lost in accuracy and discipline.

Type	Size	Move	Close Combat	Target	Save	Kill	Traits
Unteroffizier	1	4"	D6	4+	6+	6+	-
Soldier	1	4"	D6	4+	6+	6+	-

Unit: Unteroffizier (unit leader) with a MP-40 SMG, 8 Soldiers with Kar-98k rifles, Soldier with an MG-42 LMG

Tank Assault: Infantry in this period used a variety of grenades or mines to assault armoured vehicles. All members of the unit are assumed to carry Anti-Tank Explosives. These may be placed on any model of Size 2 or greater with a Charge action if the target ends up within the Lethal Range of the charging unit. They will explode at the end of the charging unit's next turn.

Untrained: This unit will lose the Auto trait on any weapons it has. In addition, it will always count cover as obscurement, and Artillery weapons gains a +1 bonus on all Damage Dice allocated to models in this unit.

Options: Up to 1 Soldier may be equipped with a Panzerfaust in addition to his rifle for +30 points. A Volksgrenadier Squad may have an Opel Blitz Medium Truck for +50 points.

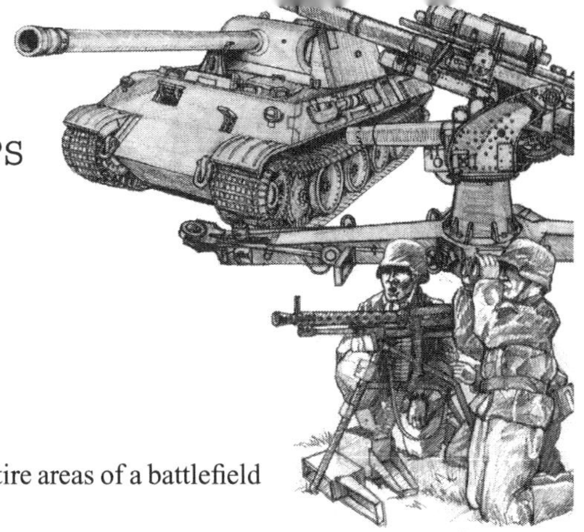

Support Assets

The following are all the Support Assets available to a Heer Grenadier force.

Grenadier MMG Support Unit - 90 points

Feared by the Allies, these machine gunners could lay down a devastating hail of fire, denying entire areas of a battlefield to the enemy.

Type	Size	Move	Close Combat	Target	Save	Kill	Traits
Unteroffizier	1	4"	D6+1	4+	6+	6+	-
Soldier	1	4"	D6+1	4+	6+	6+	-

Unit: Unteroffizier (unit leader) with a MP-40 SMG, Soldier with an MG 42 MMG, Soldier with a Kar-98k rifle

Options: A Grenadier MMG Support Unit may have a Kuebelwagen for +35 points, an SdKfz 251/1 halftrack for +80 points, or an SdKfz 250/1 Neu halftrack for +70 points.

Grenadier Panzerschreck Team - 70 points

Trained to hunt down enemy armour, these Grenadiers are the terror of Allied tanks.

Type	Size	Move	Close Combat	Target	Save	Kill	Traits
Soldier	1	4"	D6+1	4+	6+	6+	-

Unit: Soldier (unit leader) with a Panzerschreck, Soldier with a Kar-98k rifle

Options: A Grenadier Panzerschreck Team may have a Kuebelwagen for +35 points, or an SdKfz 250/1 Neu halftrack for +70 points.

Grenadier 8cm Mortar Team - 70 points

Quickly setting their mortars up during an advance, these teams can lay down heavy salvoes on very short notice.

Type	Size	Move	Close Combat	Target	Save	Kill	Traits
Unteroffizier	1	4"	D6+1	4+	6+	6+	-
Soldier	1	4"	D6+1	4+	6+	6+	-

Unit: Unteroffizier (unit leader) with a MP-40, Soldier with an 8cm Mortar (FaF), Soldier with a Kar-98k

Options: A Grenadier 8cm Mortar Team may have a Kuebelwagen for +35 points, or an SdKfz 250/1 Neu halftrack for +70 points.

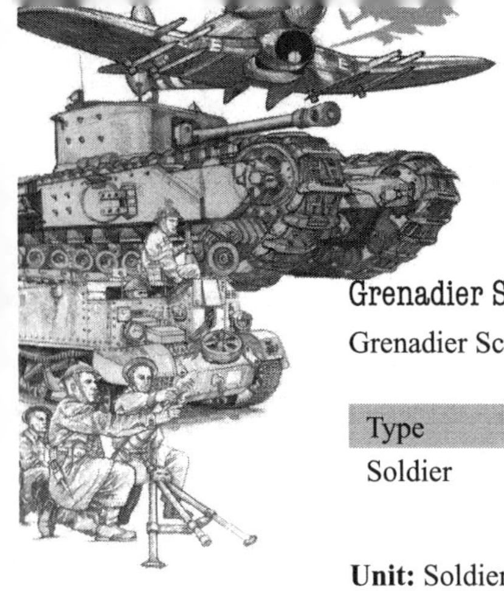

German Grenadiers

Grenadier Sniper Team - 50 points

Grenadier Scharfschützen (sniper) teams were used to pin down enemy squads or kill officers.

Type	Size	Move	Close Combat	Target	Save	Kill	Traits
Soldier	1	5"	D6+1	4+	6+	6+	-

Unit: Soldier (unit leader) with a Scoped Kar-98k rifle, Soldier with a Kar-98k rifle.

Stealthy: If a model from this unit is in Cover and did not make a Shoot action or reaction in the last turn, enemy models more than 20" away may not draw Line of Sight to it.

Ambush: This unit may be concealed in an area of terrain outside the enemy's deployment zone before the game begins. Note the location of the unit before any models are deployed on the table. You can reveal the ambush by placing the section on the table at any time after a unit (friendly or otherwise) has completed an action. However, the unit must be revealed by the end of turn three or be counted as destroyed.

Options: A Grenadier Sniper Team may have a Kuebelwagen for +35 points, or an SdKfz 250/1 Neu halftrack for +70 points. If it picks either, it may not use the Ambush rule.

Grenadier 7.5cm L46 PaK40 Anti-Tank Gun- 105 points

These anti-tank guns would be quickly set up and used to annihilate enemy armour.

Type	Size	Move	Close Combat	Target	Save	Kill	Traits
Unteroffizier	1	4"	D6+1	4+	6+	6+	-
Soldier	1	4"	D6+1	4+	6+	6+	-

Unit: Unteroffizier (unit leader) with a MP-40 SMG, Soldier with a 7.5cm L46 anti-tank gun (FaF), 2 Soldiers with Kar-98k rifles.

Crew Weapon: The 7.5cm L46 anti-tank gun can only shoot if at least one crew member is in base contact with it.

Cumbersome: A Ready action must be taken before 7.5cm L46 anti-tank gun may be moved. The Move score of all models in the unit depends on the number of models still manning the weapon.

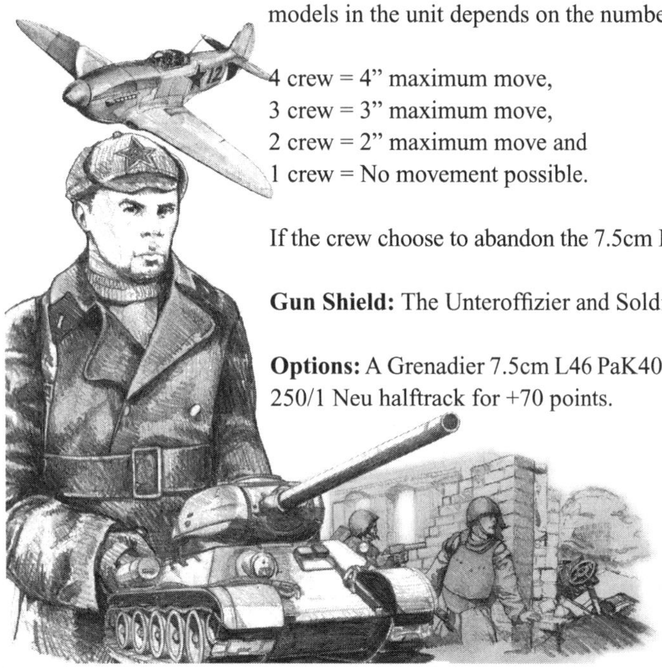

4 crew = 4" maximum move,
3 crew = 3" maximum move,
2 crew = 2" maximum move and
1 crew = No movement possible.

If the crew choose to abandon the 7.5cm L46 anti-tank gun, they may move normally.

Gun Shield: The Unteroffizier and Soldiers count as being in Medium Cover when shot from the Forward facing.

Options: A Grenadier 7.5cm L46 PaK40 Anti-Tank Gun may have an SdKfz 251/1 halftrack for +80 points, or an SdKfz 250/1 Neu halftrack for +70 points.

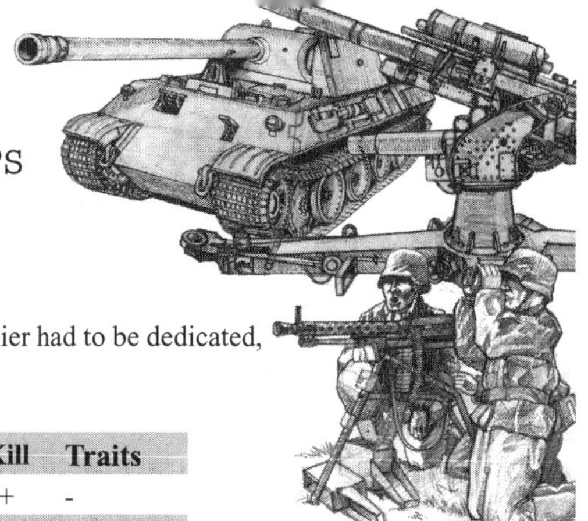

German Pionier Squad – 260 points

Whether they were building bridges, attacking enemy bunkers or disarming explosives, the Pionier had to be dedicated, highly trained and tough.

Type	Size	Move	Close Combat	Target	Armour	Kill	Traits
Unteroffizier	1	5”	D6+1	4+	6+	6+	-
Obergefreiter	1	5”	D6+1	4+	6+	6+	-
Soldier	1	5”	D6+1	4+	6+	6+	-
Goliath	1	4”	2xD10	4+	6+	6+	Retaliate

Team 1: Unteroffizier (unit leader) with a MP-40 SMG, 3 Soldiers with Kar-98k rifles, Soldier with a Flammenwerfer-41.

Team 2: Obergefreiter with a MP-40 SMG, 3 Soldiers with Kar-98k rifles, Soldier with an MG-42 LMG.

Elite Training: The unit gains a +1 bonus to their Armour Saves when in Cover. They may also re-roll their Close Combat dice.

Fearless: The unit discards the first Suppression dice it suffers every turn.

Fire Discipline: The unit need only discard one Damage Dice for each extra Fire Zone created.

Teams: The Obergefreiter may act as unit leader to Fire Team 2, splitting off to act as a separate unit.

Tank Assault: Infantry in this period used a variety of grenades or mines to assault armoured vehicles. All members of the unit are assumed to carry Anti-Tank Explosives. These may be placed on any model of Size 2 or greater with a Charge action if the target ends up within the Lethal Range of the charging unit. They will explode at the end of the charging unit’s next turn.

Goliath Tracked Mine: The Goliath tracked mine was an unmanned German-engineered demolition vehicle, carrying 75-100kg of high explosives. The vehicle was steered remotely via a joystick control box, which itself was attached to the Goliath by a triple-strand telephone cable connected to the rear of the vehicle. Each Goliath was disposable, being intended to be blown up with its target.

Declare one Soldier to be the controller of the Goliath. So long as the Controller has Line of Sight to the Goliath, he can guide it towards its target by using a Shoot action (it will move 4” for every Shoot action spent this way). The Goliath itself does not need to be in command, though it can not take actions without a controller.

If the rest of the squad is destroyed, the Goliath is also removed from the table.

Options: The Unteroffizier and Obergefreiter may swap their MP-40 SMG for a StG-44 for +7 points each. Up to 1 Soldier of either Team 1 or 2 may be equipped with a Panzerfaust in addition to his rifle for +30 points. Up to 2 Soldiers per team may swap their Kar 98k rifle for Gewehr 43 rifles for +2 points each. Any Soldier may swap their Kar 98k rifle or MP-40 for a StG-44 for +7 points each. Either team may be equipped with Improved Anti-Tank Explosives for +30 points per team. Either team may be equipped with a Goliath tracked mine for +30 points per team. Drop Team 2 for – 140 points. A Pionier Squad may be mounted in an SdKfz 251/7 halftrack for +85 points.

Late War German Armoury

Weapon	Range	Damage	Traits
7.5cm L46 Anti-Tank Gun	48”	D10+2	Lethal Zone/1“, Ready, Slow
8cm Mortar	48”	D6	Artillery, Lethal Zone/3”, Ready, Slow
Anti-Tank Explosives	Placed	D10	Lethal Zone/2“, One-Shot, Piercing/2
FG-42 LMG	24”	D6+1	Auto, Prone
Flammenwerfer-41	10”	3xD6	Flame, Slow
Gewehr 43 rifle	24”	D6	Auto
Goliath tracked mine	4”	2xD10	Lethal Zone/4“, Multihit, One-Shot, Piercing/1
Heavy Artillery	Special	2xD6+1	Artillery , Lethal Zone/4“, One-Shot
Improved Anti-Tank Explosives	Placed	2xD10+2	Lethal Zone/2“, One-Shot, Piercing/3
Kar-98k rifle	24”	D6	-
Light Artillery	Special	D6	Artillery , Lethal Zone/4“, One-Shot
MG-42 MMG	48”	3xD6+1	Auto, Prone
MG-42 LMG	48”	2xD6+1	Auto, Prone
MP-40 SMG	12”	3xD6	-
Panzerfaust	12”	D10+3	Lethal Zone/1“, Piercing/1, Slow
Panzerschreck	18”	D10+3	Lethal Zone/1“, Piercing/1, Slow
Scoped Kar-98k	36”	D6	Accurate
StG-44	18”	2xD6	Auto

7.5cm L46 Anti-Tank Gun: Capable of destroying the heaviest of Allied tanks, the L46 included a gun shield to protect the crew from enemy fire.

8cm Mortar: Extremely accurate, and backed up with a high rate of fire, this mortar was the bane of many Allied forces.

Anti-Tank Explosives: A combination of high-powered grenades and make-shift charges, these explosives give infantry a fighting chance against armoured vehicles, though their use is only advised in built-up areas.

FG-42 LMG: An automatic rifle, the FG-42 could fire single shots, but was more often used as a light support weapon.

Flammenwerfer-41: A man-portable flame-throwing system, this weapon was used to clear enemy infantry from bunkers and built-up areas.

Gewehr 43 rifle: Germany’s general issue semi-automatic rifle, the Gewehr 43 enjoyed a great amount of success, and greatly improved the efficiency of squads who used it.

Goliath tracked mine: Employed by the Wehrmacht during World War II, this caterpillar-tracked vehicle was approximately four feet long, two wide, and one tall. It was intended to be used for multiple purposes, including destroying tanks, disrupting dense infantry formations, and the demolition of buildings.

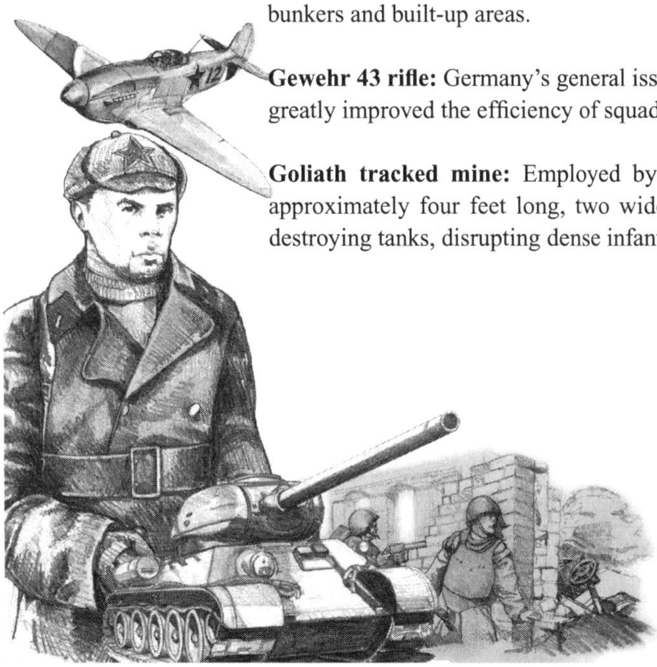

Heavy Artillery: At the call of platoon leaders, heavy artillery could literally blast infantry out of cover, and was effective against all but the heaviest of armoured vehicles.

Improved Anti-Tank Explosives: Dedicated satchel charges were used by elite units in order to immobilise and destroy enemy armour, though such attacks were extremely dangerous.

Kar-98k rifle: Simplified from pre-war designs, the Kar-98k was the standard bolt-action rifle for the Wehrmacht.

Light Artillery: Using smaller and short-ranged guns, light artillery is primarily an anti-infantry weapon, forcing enemy troops into cover and disrupting their attacks as an assault is launched.

MG-42 MMG: With a distinctive report, the MG-42 had an extremely high rate of fire, and made covered positions fearsome to assault.

MG-42 LMG: Having a lower rate of fire than its heavier cousin, this machine gun nevertheless proved a superior weapon in the squad support role.

MP-40 SMG: With a relatively low rate of fire, the MP-40 was a manageable submachine gun and was widely used as a squad leader's weapon.

Panzerfaust: A recoilless gun, the Panzerfaust was a cheap and easy to use weapon that proved deadly to enemy armour.

Panzerschreck: Though clumsy and unwieldy, the Panzerschreck was effective against any Allied tank, and was capable of blasting through even additional armoured plates.

Scoped Kar-98k rifle: The addition of a scope to the basic Kar-98k made the weapon into a solid sniper rifle.

StG-44: Arguably the first mass-produced assault rifle in the world, it was not without fault, and came too late to change the course of the war. However, this weapon was an impressive step forward.

LATE WAR GERMAN WAFFEN SS INFANTRY

The Waffen-SS was the most infamous and feared part of the German armed forces of WW II. Known as Hitler's personal bodyguard and the troops that administrated the concentration camps, the Waffen-SS also included the Combat Support Force, which served as elite troops alongside the regular army.

The Waffen-SS included many units raised from volunteers in almost all occupied territories and from amongst prisoners of war, and by the end of the war approximately sixty percent of the Waffen-SS were non-Germans. Whatever their nationality, the soldiers of the Waffen-SS tended to be highly motivated, incredibly dedicated, and very well-trained. They were a powerful and effective force.

Platoon Structure

The standard squad size for most Waffen-SS platoons is 10 soldiers, consisting of two teams of five men each. The section is led by a Unterscharführer, while the second fire team is led by a Rottenführer.

Three such sections make up a standard Waffen-SS Zug (Platoon), led by a Untersturmführer, who is in turn assisted by a Scharführer.

A platoon that has been on operational duty for some time may vary wildly from this, however. Heavy losses can result in fewer and smaller sections, while a platoon involved in heavy combat may become 'reinforced' with more soldiers, giving it more sections.

You can make any platoon with the German late war Waffen-SS Army list, from one that has just been deployed, to a grizzled set of veterans that have taken heavy losses or been reinforced.

Each unit in a platoon is considered to be an Asset of some type. You may include the following Assets in your platoon, so long as you obey the minimum and maximum units listed for each. You may take one platoon for every 1,500 points in your force.

1 Command Assets
2-4 Squad Assets
0-3 Support Assets
0-3 Armour Assets
0-1 Air Assets

For details on more Support, Armour and Air Assets see the German Vehicle and Air Assets List on page 91).

Special Rules

The following special rules apply to all German Waffen-SS platoons.

Elite Training: Every unit gains a +1 bonus to its Armour Saves when in Cover. They may also re-roll their Close Combat dice.

Fearless: Every unit discards the first Suppression dice it suffers every turn.

Fire Discipline: Every unit need only discard one Damage Dice for each extra Fire Zone created.

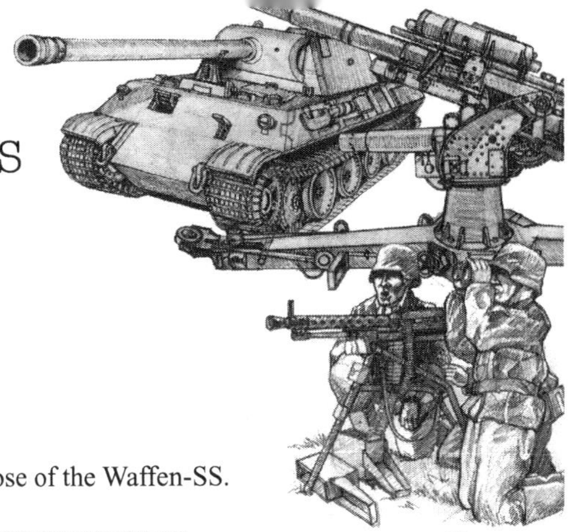

Command Assets

The following are all the Command Assets available to a German Waffen-SS force.

Waffen-SS Command Squad - 115 points

Pledging their loyalty personally to Hitler, there were few commanders more dedicated than those of the Waffen-SS.

Type	Size	Move	Close Combat	Target	Save	Kill	Traits
Untersturmführer	1	4”	D6+2	4+	6+	6+	-
Scharführer	1	4”	D6+2	4+	6+	6+	-
Soldier	1	4”	D6+2	4+	6+	6+	-

Team 1: Untersturmführer (unit leader) with a MP-40 SMG, Scharführer with a MP-40 SMG, Soldier with a Kar-98k rifle.

Team 2: Soldier with a Panzerschreck, Soldier with a Kar-98k rifle.

Fanatic: This unit automatically discards all Suppression dice.

Teams: One Soldier may act as unit leader to Team 2, splitting off to act as a separate unit.

Options: Drop Team 2 for –90 points. Add up to one additional Team 2 for +90 points. Any member of the Command squad may swap their MP-40 SMG for a StG-44 for +7 points each. The Soldier in Team 2 may swap his Kar 98k Rifle for a Gewehr 43 rifle for +2 points. A Waffen SS Command Squad may have an Opel Blitz Medium Truck for +50 points, up to 2 Kuebelwagen for +35 points each, an SdKfz 251/10 halftrack for +110 points, or up to 2 SdKfz 250/1 Neu halftracks for +70 points each.

Supporting Artillery: This unit may purchase up to 2 uses of Supporting Artillery. Calling in Supporting Artillery requires a Ready action. This is used as an Artillery attack, but the initial Impact Point must be in the unit's Line of Sight. There is no minimum range for this attack (yes, you can call it down right on your own head!). Light Artillery costs +50 points per use, while Heavy Artillery costs +100 points per use.

Squad Assets

The following are all the Squad Assets available to a German Waffen-SS force.

Waffen-SS Squad - 240 points

Regarded as an elite force, the soldiers of the Waffen-SS were among the best Germany had to offer throughout the war.

Type	Size	Move	Close Combat	Target	Save	Kill	Traits
Unterscharführer	1	4”	D6+2	4+	6+	6+	-
Rottenführer	1	4”	D6+2	4+	6+	6+	-
Soldier	1	4”	D6+2	4+	6+	6+	-

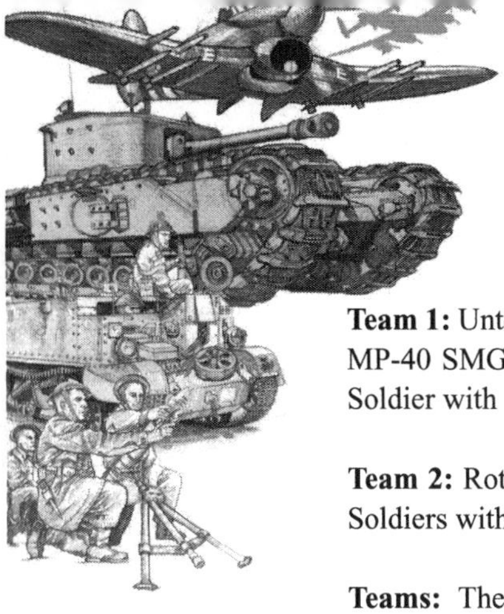

German Waffen SS

Team 1: Unterscharführer (unit leader) with a MP-40 SMG, 3 Soldiers with Kar-98k rifles, Soldier with an MG-42 LMG.

Team 2: Rottenführer with a MP-40 SMG, 4 Soldiers with Kar-98k rifles.

Teams: The Rottenführer may act as unit leader to Team 2, splitting off to act as a separate unit.

Tank Assault: Infantry in this period used a variety of grenades or mines to assault armoured vehicles. All members of the unit are assumed to carry Anti-Tank Explosives. These may be placed on any model of Size 2 or greater with a Charge action if the target ends up within the Lethal Range of the charging unit. They will explode at the end of the charging unit's next turn.

Options: Any member of either Team 1 or 2 may swap their MP-40 SMG or Kar-98k rifle for a StG-44 for +7 points each. Up to 1 Soldier of either Team 1 or 2 may be equipped with a Panzerfaust in addition to his rifle for +30 points. Any member of either Team 1 or 2 may swap their MP-40 SMG or Kar-98k rifle for a Gewehr 43 rifle for +2 points each. Up to 1 Soldier of Team 2 may be equipped with an MG-42 LMG for +40 points. Drop Team 2 for – 100 points. A Waffen-SS Squad may have an Opel Blitz Medium Truck for +50 points, an SdKfz 251/1 halftrack for +80 points or 2 SdKfz 250/1 Neu halftracks for +70 points each.

Support Assets

The following are all the Support Assets available to a German Waffen-SS force.

Waffen-SS MMG Support Unit - 110 points

With a '42' waiting for any attack, the Allies will be cautious when launching assaults against the Waffen-SS.

Type	Size	Move	Close Combat	Target	Save	Kill	Traits
Unterscharführer	1	4"	D6+2	4+	6+	6+	-
Soldier	1	4"	D6+2	4+	6+	6+	-

Unit: Unterscharführer (unit leader) with a MP-40 SMG, Soldier with an MG 42 MMG, Soldier with a Kar-98k rifle.

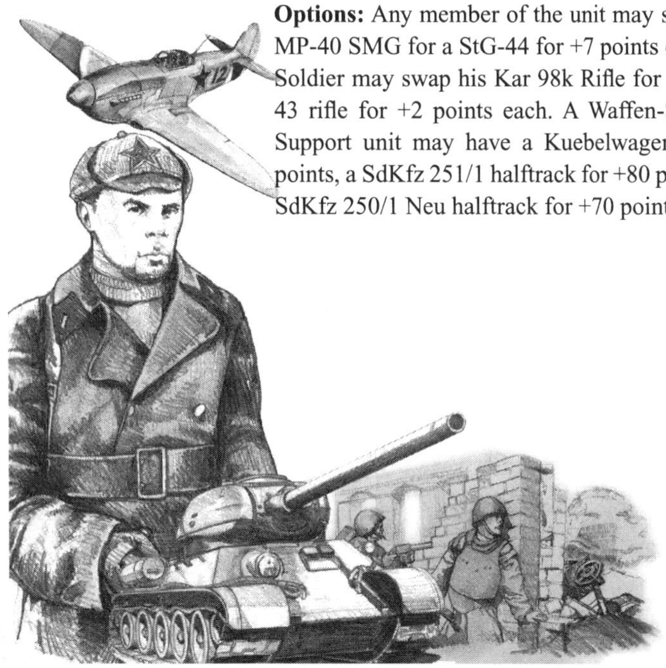

Options: Any member of the unit may swap their MP-40 SMG for a StG-44 for +7 points each. One Soldier may swap his Kar 98k Rifle for a Gewehr 43 rifle for +2 points each. A Waffen-SS MMG Support unit may have a Kuebelwagen for +35 points, a SdKfz 251/1 halftrack for +80 points, or a SdKfz 250/1 Neu halftrack for +70 points.

Waffen-SS Sniper Team - 70 points

Adept at pinning down enemy squads, the Waffen-SS had superb marksmen operating in these teams.

Type	Size	Move	Close Combat	Target	Armour	Kill	Traits
Soldier	1	5"	D6+2	4+	6+	6+	-

Unit: Soldier (unit leader) with a Scoped Kar-98k, Soldier with a Kar-98k rifle

Stealthy: If a model from this unit is in Cover and did not make a Shoot action or reaction in the last turn, enemy models more than 20" away may not draw Line of Sight to it.

Ambush: This unit may be concealed in an area of terrain outside the enemy's deployment zone before the game begins. Note the location of the unit before any models are deployed on the table. You can reveal the ambush by placing the section on the table at any time after a unit (friendly or otherwise) has completed an action. However, the unit must be revealed by the end of turn three or be counted as destroyed.

Fanatic: This unit automatically discards all Suppression dice.

Options: A Waffen-SS Sniper Team may have a Kuebelwagen for +35 points, or a SdKfz 250/1 Neu halftrack for +70 points. If it picks either, it may not use the Ambush rule.

Waffen-SS Panzerschreck Team - 85 points

Trained to seek out the weak points on enemy armour, this team can prove devastating to an enemy advance.

Type	Size	Move	Close Combat	Target	Save	Kill	Traits
Soldier	1	4"	D6+2	4+	6+	6+	-

Unit: Soldier (unit leader) Gunner with a Panzerschreck, Soldier with a Kar-98k rifle

Options: A Waffen-SS Panzerschreck Team may have a Kuebelwagen for +35 points, or a SdKfz 250/1 Neu halftrack for +70 points.

German Waffen SS

Waffen-SS 8cm Mortar Team - 95 points

Taking position on the rear line or flank of an advancing Waffen-SS force, this team can cause terrible disruption to an enemy, limiting their ability to respond.

Type	Size	Move	Close Combat	Target	Save	Kill	Traits
Unterscharführer	1	4"	D6+2	4+	6+	6+	-
Soldier	1	4"	D6+2	4+	6+	6+	-

Unit: Unterscharführer (unit leader) with a MP-40, Soldier with an 8cm Mortar (FaF), Soldier with a Kar-98k rifle.

Options: A Waffen-SS Mortar Team may have a Kuebelwagen for +35 points, or a SdKfz 250/1 Neu halftrack for +70 points.

Waffen-SS 7.5cm L46 PaK40 AT Gun - 130 points

When attached to a platoon, this team grants a very powerful anti-tank capability that can completely stall an advance.

Type	Size	Move	Close Combat	Target	Save	Kill	Traits
Unterscharführer	1	4"	D6+2	4+	6+	6+	-
Soldier	1	4"	D6+2	4+	6+	6+	-

Unit: Unterscharführer with a MP-40 SMG, Soldier with a 7.5cm L46 anti-tank gun (FaF), 2 Soldiers with Kar-98k rifles.

Crew Weapon: The 7.5cm L46 anti-tank gun can only shoot if at least one crew member is in base contact with it.

Cumbersome: A Ready action must be taken before 7.5cm L46 anti-tank gun may be moved. The Move score of all models in the unit depends on the number of models still manning the weapon.

4 crew = 4" maximum move,
3 crew = 3" maximum move,
2 crew = 2" maximum move and
1 crew = No movement possible.

If the crew choose to abandon the 7.5cm L46 anti-tank gun, they may move normally.

Gun Shield: The Unteroffizier and Soldiers count as being in Medium Cover when shot from the Forward facing.

Options: The Unterscharführer may swap his MP-40 SMG for a StG-44 for +7 points. One Soldier may swap his Kar 98k Rifle for a Gewehr 43 rifle for +2 points each. A 7.5cm L46 PaK40 AT Gun Team may have a SdKfz 251/1 halftrack for +80 points, or a SdKfz 250/1 Neu halftrack for +70 points.

Waffen-SS Pionier Squad - 245 points

Recruited from the best the Waffen-SS had to offer, the Pioniers were extremely well trained and motivated.

Type	Size	Move	Close Combat	Target	Armour	Kill	Traits
Unteroffizier	1	5”	D6+1	4+	6+	6+	-
Obergefreiter	1	5”	D6+1	4+	6+	6+	-
Soldier	1	5”	D6+1	4+	6+	6+	-
Goliath	1	4”	2xD10	4+	6+	6+	Retaliate

Team 1: Unteroffizier (unit leader) with a MP-40 SMG, 3 Soldiers with Kar-98k rifles, Soldier with a Flammenwerfer-41

Team 2: Obergefreiter with a MP-40 SMG, 3 Soldiers with Kar-98k rifles, Soldier with an MG-42 LMG.

Teams: The Obergefreiter may act as unit leader to Team 2, splitting off to act as a separate unit.

Tank Assault: Infantry in this period used a variety of grenades or mines to assault armoured vehicles. All members of the unit are assumed to carry Anti-Tank Explosives. These may be placed on any model of Size 2 or greater with a Charge action if the target ends up within the Lethal Range of the charging unit. They will explode at the end of the charging unit’s next turn.

Goliath Tracked Mine: The Goliath tracked mine was an unmanned German-engineered demolition vehicle, carrying 75-100kg of high explosives. The vehicle was steered remotely via a joystick control box, which itself was attached to the Goliath by a triple-strand telephone cable connected to the rear of the vehicle. Each Goliath was disposable, being intended to be blown up with its target.

Declare one Soldier to be the controller of the Goliath. So long as the Controller has Line of Sight to the Goliath, he can guide it towards its target by using a Shoot action (it will move 4” for every Shoot action spent this way). The Goliath itself does not need to be in command, though it can not take actions without a controller.

If the rest of the squad is destroyed, the Goliath is also removed from the table.

Options: The Unteroffizier and Obergefreiter may swap their MP-40 SMG for a StG-44 for +7 points each. Up to 1 Soldier of either Team 1 or 2 may be equipped with a Panzerfaust in addition to his rifle for +30 points. Up to 2 Soldiers per team may swap their Kar 98k Rifle for Gewehr 43 rifles for +2 points each. Any Soldier may swap their Kar 98k Rifle or Mp-40 for a StG-44 for +7 points each. Either team may be equipped with Improved Anti-Tank Explosives for +30 points per team. Either team may be equipped with a Goliath tracked mine for +30 points per team. Drop Team 2 for -140 points. The whole unit may be parachute deployed. A Waffen-SS Pionier Squad may be mounted in a SdKfz 251/7 halftrack for +85 points.

Late War German Armoury

Weapon	Range	Damage	Traits
7.5cm L46 Anti-Tank Gun	48"	D10+2	Lethal Zone/1", Ready, Slow
8cm Mortar	48"	D6	Artillery, Lethal Zone/3", Ready, Slow
Anti-Tank Explosives	Placed	D10	Lethal Zone/2", One-Shot, Piercing/2
FG-42 LMG	24"	D6+1	Auto, Prone
Flammenwerfer-41	10"	3xD6	Flame, Slow
Gewehr 43 rifle	24"	D6	Auto
Goliath tracked mine	4"	2xD10	Lethal Zone/4", Multihit, One-Shot, Piercing/1
Heavy Artillery	Special	2xD6+1	Artillery , Lethal Zone/4", One-Shot
Improved Anti-Tank Explosives	Placed	2xD10+2	Lethal Zone/2", One-Shot, Piercing/3
Kar-98k rifle	24"	D6	-
Light Artillery	Special	D6	Artillery , Lethal Zone/4", One-Shot
MG-42 MMG	48"	3xD6+1	Auto, Prone
MG-42 LMG	48"	2xD6+1	Auto, Prone
MP-40 SMG	12"	3xD6	-
Panzerfaust	12"	D10+3	Lethal Zone/1", Piercing/1, Slow
Panzerschreck	18"	D10+3	Lethal Zone/1", Piercing/1, Slow
Scoped Kar-98k	36"	D6	Accurate
StG-44	18"	2xD6	Auto

7.5cm L46 Anti-Tank Gun: Capable of destroying the heaviest of Allied tanks, the L46 included a gun shield to protect the crew from enemy fire.

8cm Mortar: Extremely accurate, and backed up with a high rate of fire, this mortar was the bane of many Allied forces.

Anti-Tank Explosives: A combination of high-powered grenades and make-shift charges, these explosives give infantry a fighting chance against armoured vehicles, though their use is only advised in built-up areas.

FG-42 LMG: An automatic rifle, the FG-42 could fire single shots, but was more often used as a light support weapon.

Flammenwerfer-41: A man-portable flame-throwing system, this weapon was used to clear enemy infantry from bunkers and built-up areas.

Gewehr 43 rifle: Germany's general issue semi-automatic rifle, the Gewehr 43 enjoyed a great amount of success, and greatly improved the efficiency of squads who used it.

Goliath tracked mine: Employed by the Wehrmacht during World War II, this caterpillar-tracked vehicle was approximately four feet long, two wide, and one tall. It was intended to be used for multiple purposes, including destroying tanks, disrupting dense infantry formations, and the demolition of buildings.

Heavy Artillery: At the call of platoon leaders, heavy artillery could literally blast infantry out of cover, and was effective against all but the heaviest of armoured vehicles.

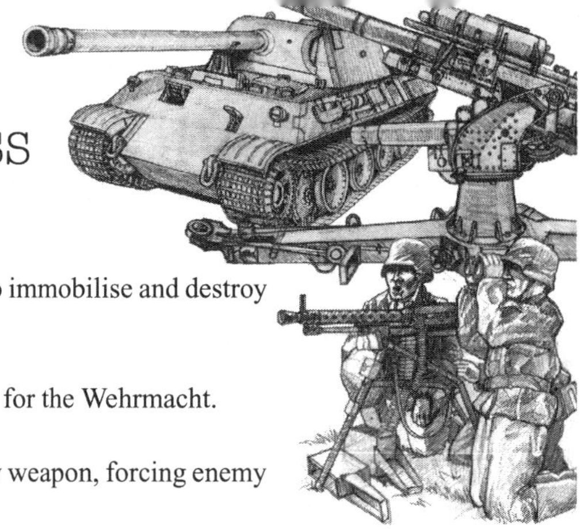

Improved Anti-Tank Explosives: Dedicated satchel charges were used by elite units in order to immobilise and destroy enemy armour, though such attacks were extremely dangerous.

Kar-98k rifle: Simplified from pre-war designs, the Kar-98k was the standard bolt-action rifle for the Wehrmacht.

Light Artillery: Using smaller and short-ranged guns, light artillery is primarily an anti-infantry weapon, forcing enemy troops into cover and disrupting their attacks as an assault is launched.

MG-42 MMG: With a distinctive report, the MG-42 had an extremely high rate of fire, and made covered positions fearsome to assault.

MG-42 LMG: Having a lower rate of fire than its heavier cousin, this machine gun nevertheless proved a superior weapon in the squad support role.

MP-40 SMG: With a relatively low rate of fire, the MP-40 was a manageable submachine gun and was widely used as a squad leader's weapon.

Panzerfaust: A recoilless gun, the Panzerfaust was a cheap and easy to use weapon that proved deadly to enemy armour.

Panzerschreck: Though clumsy and unwieldy, the Panzerschreck was effective against any Allied tank, and was capable of blasting through even additional armoured plates.

Scoped Kar-98k rifle: The addition of a scope to the basic Kar-98k made the weapon into a solid sniper rifle.

StG-44: Arguably the first mass-produced assault rifle in the world, it was not without fault, and came too late to change the course of the war. However, this weapon was an impressive step forward.

LATE WAR GERMAN FALLSCHIRMJAGER

The first force to be used in large scale airborne operations, the Fallschirmjäger showed the Allies how effective this method of attack could be. The first of these attacks was made against Denmark, but other campaigns soon followed, including Norway and Crete. During the final phases of the war, the Fallschirmjäger were no longer trained as paratroopers, but they retained their title.

Very well equipped, the Fallschirmjäger were regarded as elite troops, and the FG-42 automatic weapon was made widely available to them.

Platoon Structure

The standard squad size for most Fallschirmjäger platoons is 10 soldiers, consisting of two teams of five men each. The section is led by a Unteroffizier, while the second fire team is led by a Obergefreiter.

Three such sections make up a standard Fallschirmjäger Zug (platoon), led by a Leutnant, who is in turn assisted by a Unterfeldwebel.

A platoon that has been on operational duty for some time may vary wildly from this, however. Heavy losses can result in fewer and smaller sections, while a platoon involved in heavy combat may become 'reinforced' with more soldiers, giving it more sections.

You can make any platoon with the late war Fallschirmjäger Army list, from one that has just been deployed, to a grizzled set of veterans that have taken heavy losses or been reinforced.

Each unit in a platoon is considered to be an Asset of some type. You may include the following Assets in your platoon, so long as you obey the minimum and maximum units listed for each. You may take one platoon for every 1,500 points in your force.

0-1 Command Assets
2-4 Squad Assets
0-1 Support Asset per Squad Asset
0-1 Armour Asset per two Squad assets
0-1 Air Assets

For details on more Support, Armour and Air Assets see the German Vehicle and Air Assets List on page 91).

Special Rules

The following special rules apply to all Fallschirmjäger platoons.

Elite Training: Every unit gains a +1 bonus to their Armour Saves when in Cover. They may also re-roll their Close Combat dice.

Fearless: Every unit discards the first Suppression dice it suffers every turn.

Fire Discipline: Every unit need only discard one Damage Dice for each extra Fire Zone created.

Parachute Force Deployment: The last Fallschirmjäger deployment by parachute was during the Battle of the Bulge. It was an ill fated attempt to make a combat drop behind enemy lines under bad weather. If you want to use a parachute deployment for your force, all units eligible to take the parachute deployment option must do so (see page 38). In addition you may take two Air Assets per platoon rather than one.

Units that cannot use the parachute deployment advanced rule (such as transports or armour) may not be used in the force.

Command Assets

The following are all the Command Assets available to a Fallschirmjäger force.

Fallschirmjager Command Squad - 105 points

Trained to use their own initiative, Fallschirmjäger officers are able to quickly cope with changing battlefield conditions.

Type	Size	Move	Close Combat	Target	Save	Kill	Traits
Leutnant	1	5"	D6+1	4+	6+	6+	-
Unterfeldwebel	1	5"	D6+1	4+	6+	6+	-
Soldier	1	5"	D6+1	4+	6+	6+	-

Unit: Leutnant (unit leader) with a MP-40 SMG, Unterfeldwebel with a MP-40 SMG, 3 Soldiers with Kar-98k rifles.

Options: This unit may be parachute deployed. Any Soldier may swap their Kar-98k Rifle for a StG-44 for +7 points each. Up to 2 Soldiers may be equipped with a Panzerschreck in addition to their rifle for +40 points each. Up to 2 Soldiers may swap their Kar-98k Rifle for an FG-42 LMG for +20 points each.

Supporting Artillery: This unit may purchase up to a single use of Supporting Artillery. Calling in Supporting Artillery requires a Ready action. This is used as an Artillery attack, but the initial Impact Point must be in the unit's Line of Sight. There is no minimum range for this attack (yes, you can call it down right on your own head!). Light Artillery costs +50 points per use, while Heavy Artillery costs +100 points per use.

Squad Assets

The following are all the Squad Assets available to a Fallschirmjäger force.

Fallschirmjager Squad - 270 points

A fully equipped Fallschirmjäger squad was a dangerous enemy to every Allied troop. They were feared and respected on all fronts.

Type	Size	Move	Close Combat	Target	Armour	Kill	Traits
Unteroffizier	1	5"	D6+1	4+	6+	6+	-
Obergefreiter	1	5"	D6+1	4+	6+	6+	-
Soldier	1	5"	D6+1	4+	6+	6+	-

Team 1: Unteroffizier (unit leader) with a MP-40 SMG, 3 Soldiers with Kar-98k rifles, Soldier with an MG-42 LMG.

Team 2: Obergefreiter with a MP-40 SMG, 3 Soldiers with Kar-98k rifles, 1 Soldier with an MG-42 LMG.

Teams: The Obergefreiter may act as unit leader to Team 2, splitting off to act as a separate unit.

Tank Assault: Infantry in this period used a variety of grenades or mines to assault armoured vehicles. All members of the unit are assumed to carry Anti-Tank Explosives. These may be placed on any model of Size 2 or greater with a Charge action if the target ends up within the Lethal Range of the charging unit. They will explode at the end of the charging unit's next turn.

Options: This unit may be parachute deployed. Up to 1 Soldier of either Team 1 or 2 may be equipped with a Panzerfaust in addition to his rifle for +30 points. Up to 1 Soldier of either Team 1 or 2 may swap his Kar-98k Rifle for an FG-42 LMG for +20 points each. Any Soldier may swap their MP-40 or Kar-98k Rifle for a StG-44 for +7 points each. Drop Team 2 for – 140 points.

Support Assets

The following are all the Support Assets available to a Fallschirmjäger force.

Fallschirmjager 8cm Mortar Team – 90 points

Fallschirmjäger Kampfgruppen often did not have the same support as the normal infantry. Mortar Teams were used to overcome this shortage.

Type	Size	Move	Close Combat	Target	Armour	Kill	Traits
Unteroffizier	1	4"	D6+1	4+	6+	6+	-
Soldier	1	4"	D6+1	4+	6+	6+	-

Unit: Unteroffizier (unit leader) with a MP-40, Soldier with an 8cm Mortar (FaF), Soldier with Kar-98k.

Options: This unit may be parachute deployed

Fallschirmjager Sniper Team – 65 points

Fallschirmjäger Sniper teams were used to pin down enemy squads or take out enemy officers.

Type	Size	Move	Close Combat	Target	Armour	Kill	Traits
Soldier	1	5"	D6+1	4+	6+	6+	-

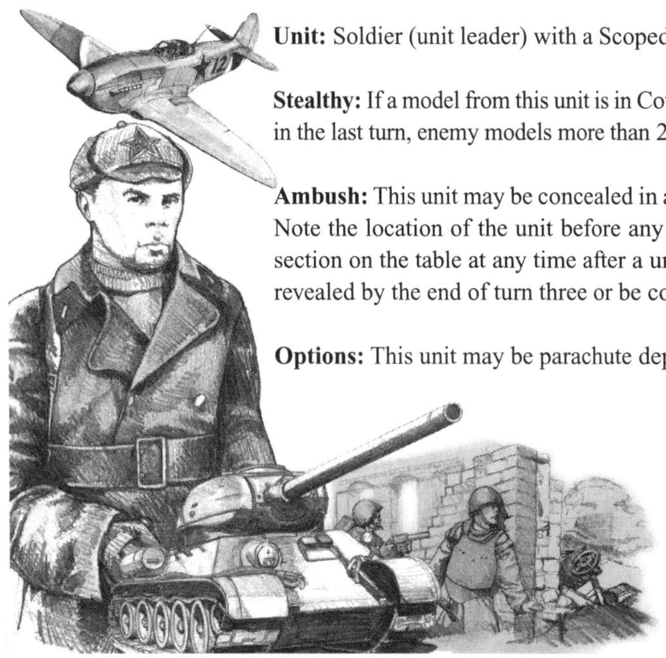

Unit: Soldier (unit leader) with a Scoped Kar-98k, Soldier with a Kar-98k.

Stealthy: If a model from this unit is in Cover and did not make a Shoot action or reaction in the last turn, enemy models more than 20" away may not draw Line of Sight to it.

Ambush: This unit may be concealed in an area of terrain outside the enemy's deployment zone before the game begins. Note the location of the unit before any models are deployed on the table. You can reveal the ambush by placing the section on the table at any time after a unit (friendly or otherwise) has completed an action. However, the unit must be revealed by the end of turn three or be counted as destroyed.

Options: This unit may be parachute deployed.

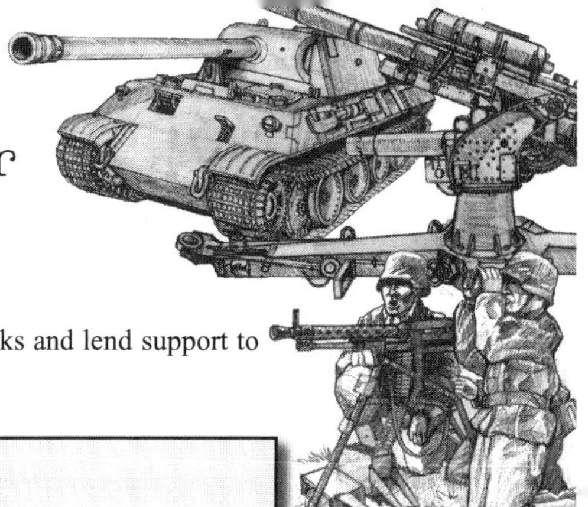

German Fallschirmjager

Fallschirmjager MMG Support Team - 110 points

Once deployed, the Fallschirmjäger could rely on these teams to cut down enemy counterattacks and lend support to their own assaults.

Type	Size	Move	Close Combat	Target	Armour	Kill	Traits
Unteroffizier	1	4"	D6+1	4+	6+	6+	-
Soldier	1	4"	D6+1	4+	6+	6+	-

Unit: Unteroffizier (unit leader) with a MP-40 SMG, Soldier with an MG 42 MMG, Soldier with a Kar 98k rifle.

Options: This unit may be parachute deployed.

Fallschirmjager Panzerschreck Team - 80 points

With Fallschirmjäger teams roving supposedly safe areas with Panzerschrecks, there were few places an Allied tank crew could relax.

Type	Size	Move	Close Combat	Target	Armour	Kill	Traits
Soldier	1	4"	D6+1	4+	6+	6+	-

Unit: Soldier (unit leader) with Panzerschreck, Soldier with Kar-98k rifle.

Options: This unit may be parachute deployed.

Luftlandepionier Squad - 245 points

Whenever the Fallschirmjäger faced a tough or unusual target, the Luftlandepioniers could be counted on to bring a hammer blow to the problem.

Type	Size	Move	Close Combat	Target	Armour	Kill	Traits
Unteroffizier	1	5"	D6+1	4+	6+	6+	-
Obergefreiter	1	5"	D6+1	4+	6+	6+	-
Soldier	1	5"	D6+1	4+	6+	6+	-
Goliath	1	4"	2xD10	4+	6+	6+	Retaliate

Team 1: Unteroffizier (unit leader) with a MP-40 SMG, 3 Soldiers with Kar-98k rifles, Soldier with a Flammenwerfer 41.

Team 2: Obergefreiter with a MP-40 SMG, 3 Soldiers with Kar-98k rifles, 1 Soldier with an MG-42 LMG.

Teams: The Obergefreiter may act as unit leader to Fire Team 2, splitting off to act as a separate unit.

Tank Assault: Infantry in this period used a variety of grenades or mines to assault armoured vehicles. All members of the unit are assumed to carry Anti-Tank Explosives. These may be placed on any model of Size 2 or greater with a Charge action if the target ends up within the Lethal Range of the charging unit. They will explode at the end of the charging unit's next turn.

Goliath Tracked Mine: The Goliath tracked mine was an unmanned German-engineered demolition vehicle, carrying 75-100kg of high explosives. The vehicle was steered remotely via a joystick control box, which itself was attached to the Goliath by a triple-strand telephone cable connected to the rear of the vehicle. Each Goliath was disposable, being intended to be blown up with its target.

Declare one Soldier to be the controller of the Goliath. So long as the Controller has Line of Sight to the Goliath, he can guide it towards its target by using a Shoot action (it will move 4" for every Shoot action spent this way). The Goliath itself does not need to be in command, though it can not take actions without a controller.

If the rest of the squad is destroyed, the Goliath is also removed from the table.

Options: The Unteroffizier and Obergefreiter may swap their MP-40 SMG for a StG-44 for +7 points each. Up to 1 Soldier of either Team 1 or 2 may be equipped with a Panzerfaust in addition to his rifle for +30 points. Up to 2 Soldiers per team may swap their Kar 98k Rifle for Gewehr 43 rifles for +2 points each. Any Soldier may swap his Kar 98k Rifle for a StG-44 for +7 points each. Either team may be equipped with Improved Anti-Tank Explosives for +30 points per team. Either team may be equipped with a Goliath tracked mine for +30 points per team. Drop Team 2 for – 140 points. This unit may be parachute deployed. A Luftlandepionier Squad may be mounted in a SdKfz 251/7 halftrack for +85points but may not be parachute deployed.

Fallschirmjager 7.5cm L46 PaK40 AT Gun – 130 points

Though unavailable during the early stages of a drop, these anti-tank guns would be quickly set up after a beachhead was created, and used to annihilate enemy armour.

Type	Size	Move	Close Combat	Target	Armour	Kill	Traits
Unteroffizier	1	4"	D6+1	4+	6+	6+	-
Soldier	1	4"	D6+1	4+	6+	6+	-

Unit: Unteroffizier (unit leader) with a MP-40 SMG, Soldier with a 7.5cm L46 anti-tank gun (FaF), 2 Soldiers with Kar-98k rifles.

Crew Weapon: The 7.5cm L46 anti-tank gun can only shoot if at least one crew member is in base contact with it.

Cumbersome: A Ready action must be taken before 7.5cm L46 anti-tank gun may be moved. The Move score of all models in the unit depends on the number of models still manning the weapon.

4 crew = 4" maximum move,
3 crew = 3" maximum move,
2 crew = 2" maximum move and
1 crew = No movement possible.

If the crew choose to abandon the 7.5cm L46 anti-tank gun, they may move normally.

Gun Shield: The Unteroffizier and Soldiers count as being in Medium Cover when shot from the Forward facing.

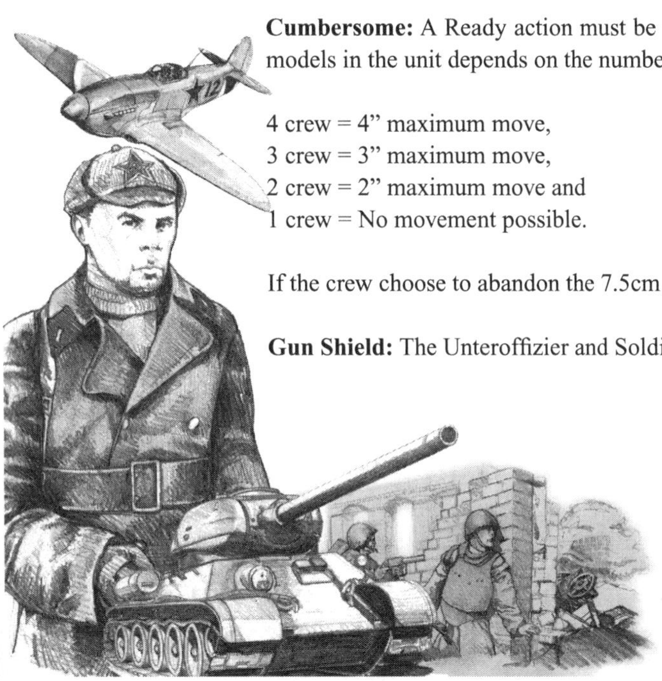

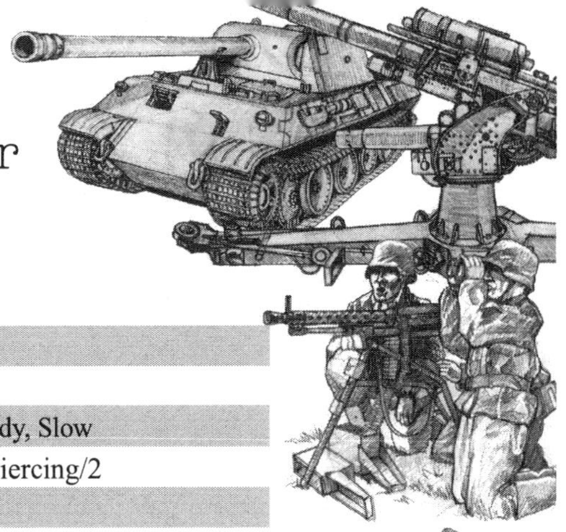

Late War German Armoury

Weapon	Range	Damage	Traits
7.5cm L46 Anti-Tank Gun	48"	D10+2	Lethal Zone/1", Ready, Slow
8cm Mortar	48"	D6	Artillery, Lethal Zone/3", Ready, Slow
Anti-Tank Explosives	Placed	2xD10+2	Lethal Zone/2", One-Shot, Piercing/2
FG-42 LMG	24"	D6+1	Auto, Prone
Flammenwerfer-41	10"	3xD6	Flame, Slow
Gewehr 43 rifle	24"	D6	Auto
Goliath tracked mine	4"	2xD10	Lethal Zone/4", Multihit, One-Shot, Piercing/1
Heavy Artillery	Special	2xD6+1	Artillery , Lethal Zone/4", One-Shot
Improved Anti-Tank Explosives	Placed	2xD10+2	Lethal Zone/2", One-Shot, Piercing/3
Kar-98k rifle	24"	D6	-
Light Artillery	Special	D6	Artillery , Lethal Zone/4", One-Shot
MG-42 MMG	48"	3xD6+1	Auto, Prone
MG-42 LMG	48"	2xD6+1	Auto, Prone
MP-40 SMG	12"	3xD6	-
Panzerfaust	12"	D10+3	Lethal Zone/1", Piercing/1, Slow
Panzerschreck	18"	D10+3	Lethal Zone/1", Piercing/1, Slow
Scoped Kar-98k	36"	D6	Accurate
StG-44	18"	2xD6	Auto

7.5cm L46 Anti-Tank Gun: Capable of destroying the heaviest of Allied tanks, the L46 included a gun shield to protect the crew from enemy fire.

8cm Mortar: Extremely accurate, and backed up with a high rate of fire, this mortar was the bane of many Allied forces.

Anti-Tank Explosives: A combination of high-powered grenades and make-shift charges, these explosives give infantry a fighting chance against armoured vehicles, though their use is only advised in built-up areas.

FG-42 LMG: An automatic rifle, the FG-42 could fire single shots, but was more often used as a light support weapon.

Flammenwerfer-41: A man-portable flame-throwing system, this weapon was used to clear enemy infantry from bunkers and built-up areas.

Gewehr 43 rifle: Germany's general issue semi-automatic rifle, the Gewehr 43 enjoyed a great amount of success, and greatly improved the efficiency of squads who used it.

Goliath tracked mine: Employed by the Wehrmacht during World War II, this caterpillar-tracked vehicle was approximately four feet long, two wide, and one tall. It was intended to be used for multiple purposes, including destroying tanks, disrupting dense infantry formations, and the demolition of buildings.

Heavy Artillery: At the call of platoon leaders, heavy artillery could literally blast infantry out of cover, and was effective against all but the heaviest of armoured vehicles.

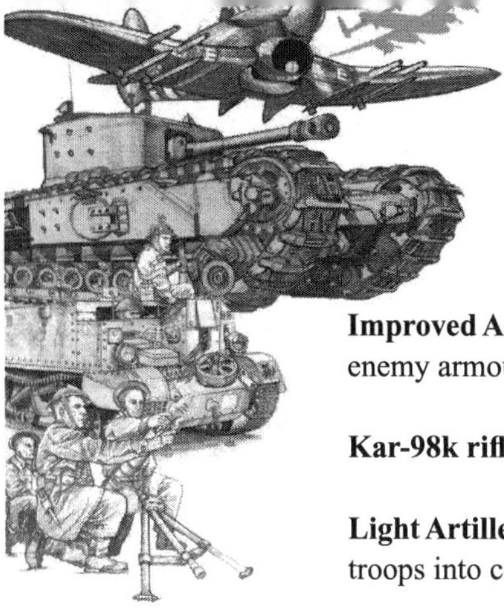

German Fallschirmjager

Improved Anti-Tank Explosives: Dedicated satchel charges were used by elite units in order to immobilise and destroy enemy armour, though such attacks were extremely dangerous.

Kar-98k rifle: Simplified from pre-war designs, the Kar-98k was the standard bolt-action rifle for the Wehrmacht.

Light Artillery: Using smaller and short-ranged guns, light artillery is primarily an anti-infantry weapon, forcing enemy troops into cover and disrupting their attacks as an assault is launched.

MG-42 MMG: With a distinctive report, the MG-42 had an extremely high rate of fire, and made covered positions fearsome to assault.

MG-42 LMG: Having a lower rate of fire than its heavier cousin, this machine gun nevertheless proved a superior weapon in the squad support role.

MP-40 SMG: With a relatively low rate of fire, the MP-40 was a manageable submachine gun and was widely used as a squad leader's weapon.

Panzerfaust: A recoilless gun, the Panzerfaust was a cheap and easy to use weapon that proved deadly to enemy armour.

Panzerschreck: Though clumsy and unwieldy, the Panzerschreck was effective against any Allied tank, and was capable of blasting through even additional armoured plates.

Scoped Kar-98k rifle: The addition of a scope to the basic Kar-98k made the weapon into a solid sniper rifle.

StG-44: Arguably the first mass-produced assault rifle in the world, it was not without fault, and came too late to change the course of the war. However, this weapon was an impressive step forward.

LATE WAR GERMAN VEHICLES AND AIR ASSETS

The following Assets may be used by any Late War German force.

Support Assets

The following are all the Support Assets available to a German force.

8.8cm FlaK 35/36 Heavy AA/AT Dual Purpose Gun – 140 points

Packing a heavy punch the '88' was pressed into service as an anti-tank gun early in the war and still proved effective at stopping even the heaviest of tanks late in the war.

Type	Size	Move	Close Combat	Target	Armour	Kill	Traits
Commander	1	4"	D6	4+	6+	6+	Immovable
Soldier	1	4"	D6	4+	6+	6+	Immovable

Unit: Commander (unit leader), Soldier with an 8.8cm FlaK Heavy AA/AT Gun (FaF), 4 Soldiers.

Crew Weapon: The 8.8cm FlaK Heavy AA/AT gun can only shoot if at least one crew member is in base contact with it.

Dual Purpose Weapon: Dual Purpose Guns fulfilled two roles, designed as a compromise between the heavy main armament of an anti-tank gun and dedicated anti-aircraft gun.

Before the battle, the player fielding this unit must decide whether it will be used in the anti-tank or anti-aircraft roles. If used as an anti-aircraft gun, the 8.8cm FlaK Heavy AA/AT gun gains the AA-Only trait. To change role during the battle, two consecutive Ready actions must be performed.

Gun Shield: The Unteroffizier and Soldiers count as being in Medium Cover when shot at from the Forward facing.

Fearless: This unit can ignore the first Suppression dice it suffers every turn.

German Vehicles

Kübelwagen – 35 points

Based on the famous Beetle, the Kübelwagen was surprisingly good across rough terrain, and able to move small teams across the battlefield rapidly.

Type	Size	Move	Close Combat	Target	Save	Kill	Traits
Kübelwagen	2	8”	D10	6+	6+	7+	Hits/2, Lumbering, Wheels/16”

Unit: Kübelwagen.

Transport: The Kübelwagen can carry up to 5 Size points of models (each maximum Size 1).

Nebelwerfer 41 – 110 points

Nicknamed the Moaning Minnie, the Nebelwerfer was a multi-barrelled rocket launcher, towed into battle.

Type	Size	Move	Close Combat	Target	Armour	Kill	Traits
Commander	1	4”	D6	4+	6+	6+	-
Soldier	1	4”	D6	4+	6+	6+	-

Unit: Commander, Soldier with a 15cm Nebelwerfer (FaF), 2 Soldiers.

Crew Weapon: The 15cm Nebelwerfer can only shoot if at least one crew member is in base contact with it.

Cumbersome: A Ready action must be taken before 7.5cm L46 anti-tank gun may be moved. The Move score of all models in the unit depends on the number of models still manning the weapon.

4 crew alive = 4” maximum move,
3 crew alive = 3” maximum move,
2 crew alive = 2” maximum move and
no move if only 1 crew is alive.

If the crews choose to abandon the weapon they may move normally.

Fearless: The German 15cm Nebelwerfer crew unit can ignore one Suppression dice every turn.

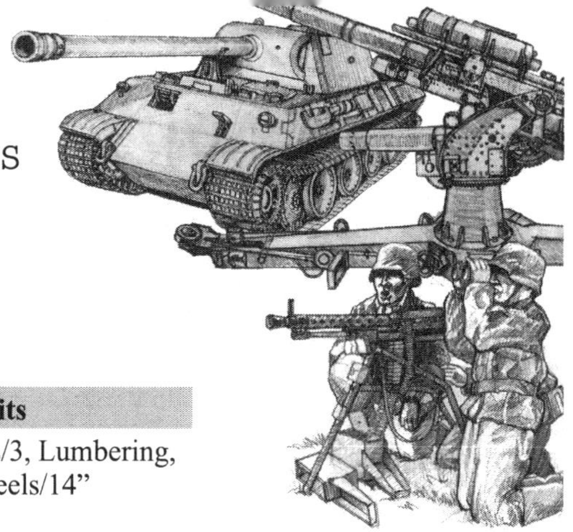

Opel Blitz Medium Truck - 50 points

This 3 ton truck was used extensively by German forces to ferry both men and equipment.

Type	Size	Move	Close Combat	Facing	Target	Save	Kill	Traits
Medium Truck	4	7"	D10		6+	6+	7+	Hits/3, Lumbering, Wheels/14"

Unit: Opel Blitz Medium Truck.

Transport: The Opel Blitz Medium Truck can carry up to 10 Size points of models (each maximum Size 1).

SdKfz 222 - 105 points

A light armoured car, the SdKfz 222 sported a powerful autocannon.

Type	Size	Move	Close Combat	Facing	Target	Save	Kill	Traits
SdKfZ 222	2	8"	2xD10	Front	7+	5+	9+	Hits/2, Lumbering, Multifire/2, Wheels/16"
				Sides	6+	6+	8+	
				Rear	6+	6+	7+	

Unit: SdKfZ 222 with a 2cm L/55 gun and an MG.

Co-Axial MG: To improve target acquisition, MGs on vehicles were often mounted co-axial to the main gun. The co-axial MG must always use the same Fire Zone as the main gun. As soon as a Fire Zone is placed within range of the co-axial MG, you may add +1 to all Damage Dice of the main gun.

Sdkfz 250/1 Neu Halftrack - 70 points

A light armoured halftrack, this vehicle had a three-ton cargo capacity, and a complete communications suite.

Type	Size	Move	Close Combat	Facing	Target	Save	Kill	Traits
SdKfz 250/1	2	6"	2xD10	Front	7+	5+	9+	Hits/2, Lumbering, Tracks/12"
				Sides	6+	6+	8+	
				Rear	6+	6+	7+	

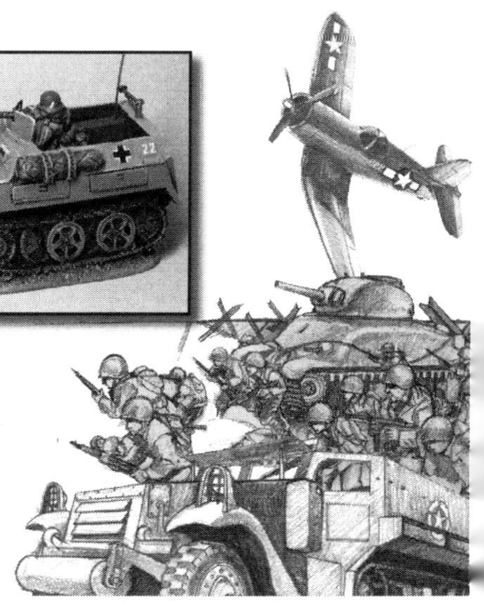

Unit: SdKfz 250/1 with a Pintle MG (FaF).

Transport: The SdKfz can carry up to 6 Size points of models (each maximum Size 1).

Options: A Pintle MG (FaA) may be added for +25 points.

German Vehicles

Sdkfz 250/7 Neu Mortar Halftrack – 100 points

The SdKfz 250/7 leichte Schützenpanzerwagen with 80mm mortar was used in a support role.

Type	Size	Move	Close Combat	Facing	Target	Save	Kill	Traits
SdKfz 250/7	2	6”	2xD10	Front	7+	5+	9+	Hits/2, Lumbering, Tracks/12”
				Sides	6+	6+	8+	
				Rear	6+	6+	7+	

Unit: SdKfz 250/7 with a Pintle MG (FaA) and an 8cm Mortar (FaF).

Sdkfz 250/8 Neu ‚Stummel‘ Halftrack – 80 points

The SdKfz 250/8 leichte Schützenpanzerwagen was a support variant armed with a 75mm L/24 gun and an MG-34.

Type	Size	Move	Close Combat	Facing	Target	Save	Kill	Traits
SdKfz 250/8	2	6”	2xD10	Front	7+	5+	9+	Hits/2, Lumbering, Tracks/12”
				Sides	6+	6+	8+	
				Rear	6+	6+	7+	

Unit: SdKfz 250/8 with a Pintle MG (FaF) and a 7.5 cm L/24 (FaF).

Options: A Pintle MG (FaA) may be added for +25 points.

Sdkfz 250/9 Neu Halftrack – 100 points

The SdKfz 250/9 leichte Schützenpanzerwagen was a reconnaissance variant with a 20mm autocannon, coaxial with an MG34 or MG42 in a low, open topped turret identical to the SdKfz-222 armoured car.

Type	Size	Move	Close Combat	Facing	Target	Save	Kill	Traits
SdKfz 250/8	2	6”	2xD10	Front	7+	5+	9+	Hits/2, Lumbering, Tracks/12”
				Sides	6+	6+	8+	
				Rear	6+	6+	7+	

Unit: SdKfz 250/9 with a 2cm L/55 gun and an MG.

Co-Axial MG: To improve target acquisition, MGs on vehicles were often mounted co-axial to the main gun. The co-axial MG must always use the same Fire Zone as the main gun. As soon as a Fire Zone is placed within range of the co-axial MG, you may add +1 to all Damage Dice of the main gun.

Options: A Pintle MG (FaA) may be added for +25 points.

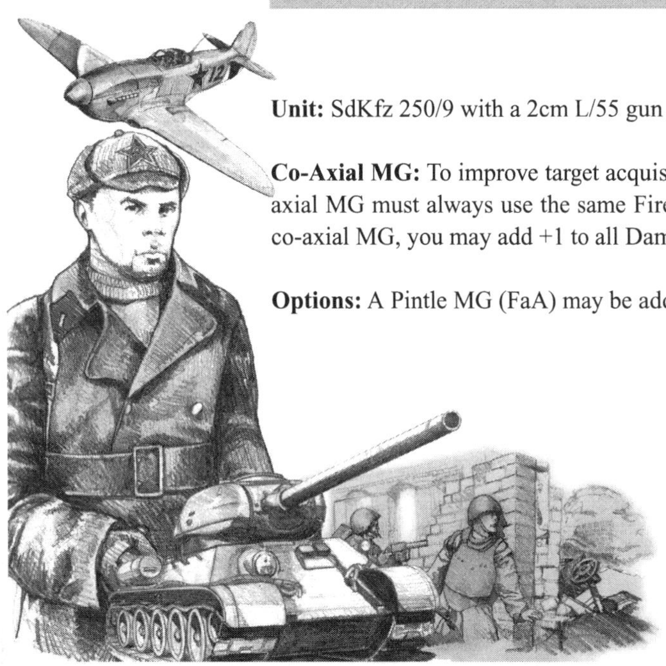

SdKfz 251/1 Halftrack - 80 points

An armoured fighting vehicle designed and first built by Nazi Germany's Hanomag company during World War II, it was intended to transport a squad of infantry to the battlefield protected from enemy small arms fire.

Type	Size	Move	Close Combat	Facing	Target	Save	Kill	Traits
SdKfz 251/1	2	5"	2xD10	Front	7+	5+	9+	Hits/2, Lumbering, Tracks/11"
				Sides	6+	6+	8+	
				Rear	6+	6+	7+	

Unit: SdKfz 251/1 with a Pintle MG (FaF).

Transport: The SdKfz 251/1 can carry up to 10 Size points of models (each maximum Size 1).

Options: A Pintle MG (FaA) may be added for +25 points.

SdKfz 251/7 Pionier Halftrack - 85 points

Designed specifically for the Pioneers, this halftrack comes complete with its own bridging equipment.

Type	Size	Move	Close Combat	Facing	Target	Save	Kill	Traits
SdKfz 251/7	2	5"	2xD10	Front	7+	5+	9+	Hits/2, Lumbering, Tracks/11"
				Sides	6+	6+	8+	
				Rear	6+	6+	7+	

Unit: 1 SdKfz 251/7 with a Pintle MG (FaF).

Transport: The SdKfz can carry up to 10 Size points of models (each maximum Size 1).

Options: A Pintle MG (FaA) may be added for +25 points.

Sdkfz 251/9 Stummel Halftrack - 80 points

Equipped with a 7.5 cm L/24 low velocity gun, this halftrack was nicknamed the Stummel (the Stump).

Type	Size	Move	Close Combat	Facing	Target	Save	Kill	Traits
SdKfz 250/8	2	5"	2xD10	Front	7+	5+	9+	Hits/2, Lumbering, Tracks/11"
				Sides	6+	6+	8+	
				Rear	6+	6+	7+	

Unit: SdKfz 251/9 with a Pintle MG (FaF) and a 7.5 cm L/24 (FaF).

Options: A Pintle MG (FaA) may be added for +25 points.

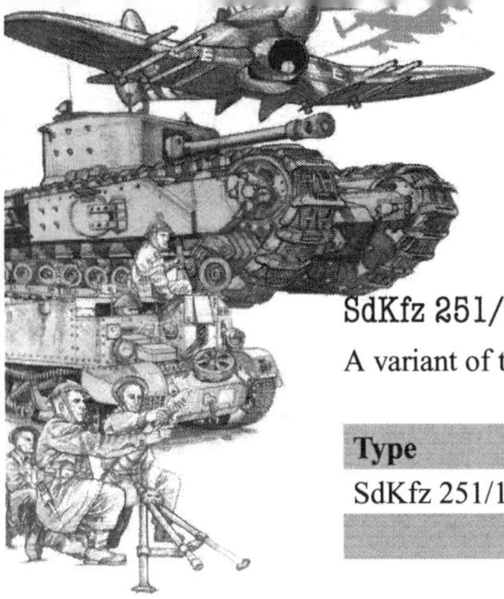

SdKfz 251/10 Halftrack - 110 points

A variant of the standard halftrack, the 251/10 was equipped with a 37 mm Pak 36 anti-tank gun.

Type	Size	Move	Close Combat	Facing	Target	Save	Kill	Traits
SdKfz 251/10	2	5”	2xD10	Front	7+	5+	9+	Hits/2, Lumbering, Tracks/11”
				Sides	6+	6+	8+	
				Rear	6+	6+	7+	

Unit: SdKfz 251/10 with a 3.7cm L45 (FaF) and a Pintle MG (FaF).

Transport: The SdKfz 251/10 can carry up to 7 Size points of models (each maximum Size 1).

Options: A Pintle MG (FaA) may be added for +25 points.

Wirbelwind Medium Anti Aircraft Tank - 185 points

The Wirbelwind was a rapid deployable anti-aircraft tank based on the widespread Panzer IV chassis.

Type	Size	Move	Close Combat	Facing	Target	Save	Kill	Traits
Wirbelwind	4	4”	4xD10	Front	8+	4+	11+	Hits/2, Lumbering, Multifire/2, Tracks/8”
				Sides	7+	5+	10+	
				Rear	7+	5+	9+	

Unit: Wirbelwind with a 2cm Quad anti-aircraft gun and an MG (FaF).

Armour Assets

The following are all the Armour Assets available to a German force.

Jagdpanzer 38 Hetzer Tankhunter – 200 points

Better armoured than the earlier Panzerjäger Marder and Nashorn, the Hetzer carried a reasonably powerful gun, and was mechanically reliable.

Type	Size	Move	Close Combat	Facing	Target	Save	Kill	Traits
Hetzer	4	4”	4xD10	Front	9+	3+	12+	Hits/2, Lumbering, Multifire/2, Tracks/8”
				Sides	8+	4+	11+	
				Rear	7+	6+	9+	

Unit: Hetzer with a 7.5cm L48 anti-tank gun (FaF) and a Pintle MG (FaF).

Jagdpanzer IV/70 Tankhunter – 240 points

A tank destroyer based on the Panzer IV chassis, this vehicle had very heavy frontal armour.

Type	Size	Move	Close Combat	Facing	Target	Save	Kill	Traits
Jagdpanzer IV/70	4	4”	4xD10	Front	9+	3+	12+	Hits/2, Lumbering, Multifire/2, Tracks/7”
				Sides	8+	4+	11+	
				Rear	8+	5+	10+	

Unit: Jagdpanzer IV/70 with a 7.5cm L70 anti-tank gun (FaF) and a Pintle MG (FaF).

Jagdpanzer V Jagdpanther Tankhunter – 400 points

The Jagdpanther (Hunting Panther) was fast and had a powerful main gun, enabling it to destroy any type of Allied tank.

Type	Size	Move	Close Combat	Facing	Target	Save	Kill	Traits
Jagdpanther	4	5”	4xD10	Front	9+	3+	13+	Hits/3, Lumbering, Multifire/2, Tracks/10”
				Sides	8+	5+	12+	
				Rear	8+	5+	11+	

Unit: Jagdpanther with an 8.8cm L71 anti-tank gun (FaF) and an MG (FaF).

German Vehicles

PzKmpf IV Ausf. H Tank - 215 points

Upgraded throughout the war, the Panzer IV was used until the very end of the war, in almost every theatre.

Type	Size	Move	Close Combat	Facing	Target	Save	Kill	Traits
Panzer IV H	4	4”	4xD10	Front	8+	4+	12+	Hits/2, Lumbering, Multifire/2, Tracks/8”
				Sides	8+	4+	12+	
				Rear	7+	5+	10+	

Unit: Panzerkampfwagen IV H with a 7.5cm L48 anti-tank gun, and two MGs.

Co-Axial MG: To improve target acquisition, MGs on vehicles were often mounted co-axial to the main gun. One MG has a co-axial mount. The co-axial MG must always use the same Fire Zone as the main gun. As soon as a Fire Zone is placed within range of the co-axial MG, you may add +1 to all Damage Dice of the main gun.

Options: Side skirts (Schürzen) may be added for +20 pts. Any failed Armour Saves from attacks on the Side facings may be re-rolled.

PzKpfw V Panther Ausf. G Tank - 420 points

Intended as a counter to the T-34, the Panther’s excellent combination of firepower, mobility and protection served as a benchmark for future tank designs.

Type	Size	Move	Close Combat	Facing	Target	Save	Kill	Traits
Panther G	4	5”	4xD10	Front	9+	3+	13+	Hits/3, Lumbering, Multifire/2, Tracks/10”
				Sides	8+	4+	12+	
				Rear	8+	5+	11+	

Unit: 1 Panther with a 7.5cm L70 anti-tank gun, a Pintle MG, an MG, and an MG (FaF).

Co-Axial MG: To improve target acquisition, MGs on vehicles were often mounted co-axial to the main gun. One MG has a co-axial mount. The co-axial MG must always use the same Fire Zone as the main gun. As soon as a Fire Zone is placed within range of the co-axial MG, you may add +1 to all Damage Dice of the main gun.

PzKpfw VI Tiger I Ausf. E Tank – 620 points

The Tiger I represented a new approach in tank design that emphasized firepower and armour at the expense of mobility.

Type	Size	Move	Close Combat	Facing	Target	Save	Kill	Traits
Tiger I E	4	4"	5xD10	Front	9+	3+	13+	Hits/4, Lumbering, Multifire/2, Tracks/8"
				Sides	9+	4+	13+	
				Rear	8+	4+	12+	

Unit: Tiger with an 8.8cm L56 anti-tank gun, an MG (FaF) and a Pintle MG.

Co-Axial MG: To improve target acquisition, MGs on vehicles were often mounted co-axial to the main gun. One MG has a co-axial mount. The co-axial MG must always use the same Fire Zone as the main gun. As soon as a Fire Zone is placed within range of the co-axial MG, you may add +1 to all Damage Dice of the main gun.

PzKpfw VI Tiger II Königstiger Tank – 800 points

The Tiger II combined the heavy armour of the Tiger I with the sloped armour of the Panther. The Königstiger has reached legendary status by the end of the war.

Type	Size	Move	Close Combat	Facing	Target	Save	Kill	Traits
Königstiger	4	4"	5xD10	Front	9+	2+	13+	Hits/4, Lumbering, Multifire/2, Tracks/8"
				Sides	9+	3+	13+	
				Rear	8+	3+	12+	

Unit: Königstiger with an 8.8cm L71 anti-tank gun, and two MGs (FaF).

Co-Axial MG: To improve target acquisition, MGs on vehicles were often mounted co-axial to the main gun. One MG has a co-axial mount. The co-axial MG must always use the same Fire Zone as the main gun. As soon as a Fire Zone is placed within range of the co-axial MG, you may add +1 to all Damage Dice of the main gun.

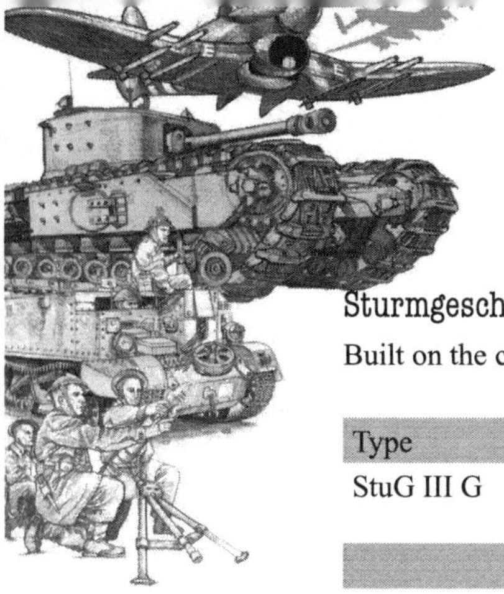

German Vehicles

SturmgeschUtz StuG III G Assault Gun - 210 points

Built on the chassis of the Panzer III tank, the StuG III was widely employed as a tank destroyer.

Type	Size	Move	Close Combat	Facing	Target	Save	Kill	Traits
StuG III G	4	4”	4xD10	Front	9+	3+	12+	Hits/2, Lumbering, Multifire/2, Tracks/8”
				Sides	8+	4+	11+	
				Rear	7+	5+	10+	

Unit: StuG III G with a 7.5cm L48 anti-tank gun (FaF) and a Pintle MG (FaF).

Air Assets

The following are all the Air Assets available to a German force.

Focke Wulf Fw 190-A8 - 180 points

The Fw 190 was well liked by its pilots, and widely regarded as superior to the front line Spitfire Mk V on its combat debut early in the war.

Type	Size	Move	Close Combat	Target	Save	Kill	Traits
FW 190 A-8	5	Loiter Cruise	4xD10	6+	5+/4+	8+	Hits/3, Multifire/4

Unit: FW 190 A-8 with four MG 151/20E 20mm cannon (FaF).

German Vehicle Armoury

Weapon	Range	Damage	Traits
15cm Nebelwerfer	60”	2xD6+1	Artillery, Lethal Zone/4”, Slow
MG 151/20E 20mm cannon	24”	1xD6+2	Piercing/1
2cm L55 gun	48”	D6+2	Lethal Zone/2”, Slow
2cm Quad AA gun	48”	4xD6+3	AA, Accurate, Slow
3.7cm L45 gun	48”	D6+3	Lethal Zone/2”, Slow
7.5cm L/24 gun	18”	D6+3	Lethal Zone/2”, Slow
7.5cm L48 anti-tank gun	48”	D10+2	Lethal Zone/1”, Piercing/2, Slow
7.5cm L70 anti-tank gun	60”	D10+4	Lethal Zone/1”, Multihit, Piercing/3, Slow
8.8cm FlaK heavy AA/AT gun	60”	D10+4	Lethal Zone/1”, Multihit, Piercing/3, Slow
8.8cm L56 anti-tank gun	60”	D10+4	Lethal Zone/1”, Multihit, Piercing/1, Slow
8.8cm L71 anti-tank gun	60”	D10+4	Lethal Zone/1”, Multihit, Piercing/3, Slow
8cm Mortar	48”	D6	Lethal Zone/3”, Artillery, Slow
MG	24”	3xD6	Auto
Pintle MG	24”	3xD6	AA, Auto

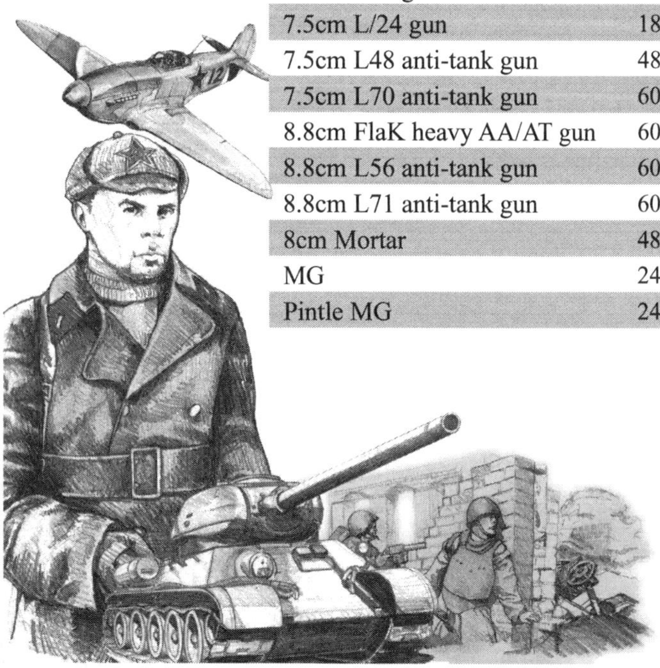

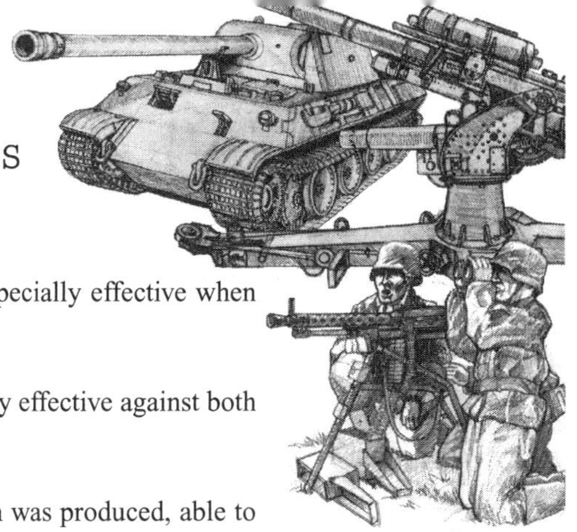

15cm Nebelwerfer: The first modern multiple rocket launcher, the 15cm Nebelwerfer was especially effective when employed against massed infantry.

MG 151/20E 20mm cannon: Used on many German aircraft of the war, this cannon was highly effective against both ground targets and other aircraft.

2cm Quad AA gun: By mounting four FlaK 38 guns together a fearsome anti-aircraft weapon was produced, able to cover whole regions of the battlefield.

3.7cm L45 gun: Originally mounted on the Panzer III, this gun was later fitted to lighter vehicles, providing them with a heavier punch.

7.5cm L/24 gun: Mounted on a variety of German vehicles, the L/24 was a solid gun, though it was out-paced by more modern designs later in the war.

7.5cm L48 anti-tank gun: Used by tank destroyers and infantry support vehicles, this gun lacked the range and penetration of its biggest rivals, but was still a potent weapon.

7.5cm L70 anti-tank gun: Incredibly accurate, and able to penetrate thick armour, this was the gun that made the Panther truly dangerous.

8.8cm FlaK heavy AA/AT gun: A dual purpose gun, this weapon could be set up to engage aircraft or armour relatively quickly.

8.8cm L56 anti-tank gun: Feared by enemy tank crews, this gun was both long-ranged and accurate, able to penetrate armour at great distance.

8.8cm L71 anti-tank gun: Used by the Tiger II and various tank destroyers, this was a large weapon that proved extremely capable.

8cm Mortar: Extremely accurate this mortar was the bane of many Allied forces.

MG: Covering a wide variety of weapons, machine guns mounted on vehicles are used to suppress enemy infantry and provide supporting fire.

Pintle MG: When seated on a pintle mount, a machine guns gains a greater field of fire and elevation, allowing it to be used to engage low-flying aircraft.

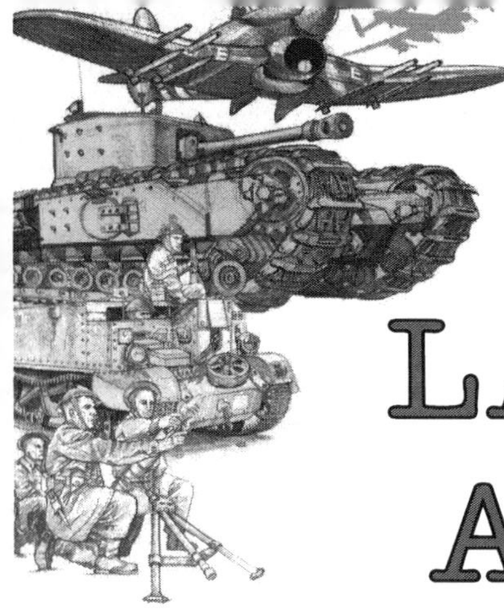

LATE WAR SOVIET ARMY INFANTRY

Paying a higher price in blood than any other nation, the Soviet Union was the main instrument responsible for the defeat of Nazi Germany. Taking crippling losses in the early period of the war, the Soviet Army soon gained a reputation for fighting against terrible odds, and making the advancing Axis forces pay for every mile of Russian territory in their own blood. Battles in Leningrad and at Kursk would become legendary in later years.

When the back of the German attack in Russia was finally broken, the Soviet Army swept forward to reclaim loss territory and marched on into Berlin. This was when their forces were at their most terrible, and at their very best.

Platoon Structure

The standard squad size for most Soviet Army infantry platoons is 9 soldiers, led by a Sergeant.

Three such sections make up a standard platoon, led by a Lieutenant, who is in turn watched over by an NKVD Officer. Motor Rifle platoons were organised in a similar fashion, with squads carried into range on trucks or halftracks provided under the Lend-Lease agreement.

A platoon that has been on operational duty for some time may vary wildly from this, however. Heavy losses can result in fewer and smaller sections, while a platoon involved in heavy combat may become 'reinforced' with more soldiers, giving it more sections.

You can make any platoon with the late war Soviet Army Infantry army list, from one that has just been deployed, to a grizzled set of veterans that have taken heavy losses or been reinforced.

Each unit in a platoon is considered to be an Asset of some type. You may include the following Assets in your platoon, so long as you obey the minimum and maximum units listed for each. You may take one platoon for every 1,500 points in your force.

0-1 Command Assets
2-4 Squad Assets
0-1 Support Asset per Squad Asset chosen
0-3 Armour Assets
0-2 Air Assets

For details on more Support, Armour and Air Assets see the Soviet Vehicle and Air Assets List on page 102).

Special Rules

Tank Desant

A combined arms tactic, riding tanks into battle, then dismounting to fight was a common ploy used by the Soviet Army. Soviet tanks were built with integrated handholds specifically for this purpose and during the winter, troops were known to travel behind tanks, towed along on skis.

Tanks capable of acting as transports are so noted in their army list descriptions.

Command Assets

The following are all the Command Assets available to a Soviet Army Infantry force.

Infantry Command Squad - 100 points

Respected by his troops, a Soviet Lieutenant was always aware of the eyes of the NKVD officer behind him.

Type	Size	Move	Close Combat	Target	Save	Kill	Traits
Lieutenant	1	4"	D6	4+	6+	6+	-
NKVD Officer	1	4"	D6	4+	6+	6+	Independent
Soldier	1	4"	D6	4+	6+	6+	-

Team 1: Lieutenant (unit leader) with a PPSh-41 SMG, Soldier with a Mosin-Nagant rifle, NKVD Officer with a PPSh-41 SMG.

Team 2: Soldier with a PTRD-41 AT rifle, Soldier with a Mosin-Nagant rifle.

Teams: One Soldier may act as unit leader to Team 2, splitting off to act as a separate unit.

NKVD Officer: The Officer can join any infantry unit and become its unit leader. Through sheer force of will the NKVD officer instils the spirit of Communism into the unit he is attached to, allowing them to discard all Suppression dice. Furthermore, he may gather up to 30 friendly Size 1 models within 12" of him to create a Human Wave. A Human Wave then moves and fights as a single unit so long as the NKVD officer is alive and they remain within 12" of him. All models in a Human Wave add +1" to their Move score.

Supporting Artillery: This unit may purchase up to 3 uses of Supporting Artillery. Calling in Supporting Artillery requires a Ready action. This is used as an Artillery attack, but the initial Impact Point must be in the unit's Line of Sight. There is no minimum range for this attack (yes, you can call it down right on your own head!). Light Artillery costs +50 points per use, while Heavy Artillery costs +100 points per use.

Options: Drop Team 2 for –40 points. Add up to one additional Team 2 for +40 points. An Infantry Command Squad may have a Soviet Medium Truck for +50 points, 1 Gaz-67 per Team for +40 points each, or an M3 halftrack for +80 points.

Squad Assets

The following are all the Squad Assets available to a Soviet Army Infantry force.

Rifle Squad - 130 points

Soviet infantry were famed for their tough and dogged approach to battle. They were often badly underestimated by their enemies.

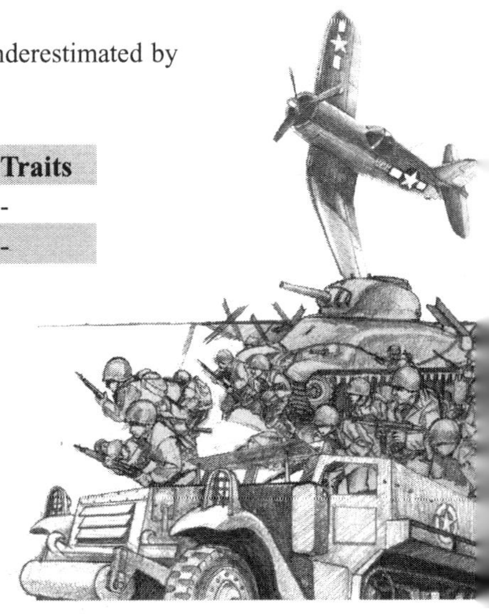

Type	Size	Move	Close Combat	Target	Save	Kill	Traits
Sergeant	1	4"	D6	4+	6+	6+	-
Soldier	1	4"	D6	4+	6+	6+	-

Unit: Sergeant (unit leader) with a PPSh-41 SMG, 7 Soldiers with Mosin-Nagant rifles, Soldier with a DP LMG.

Tank Assault: Infantry in this period used a variety of grenades or mines to assault armoured vehicles. All members of the unit are assumed to carry Anti-Tank Explosives. These may be placed on any model of Size 2 or greater with a Charge action if the target ends up within the Lethal Range of the charging unit. They will explode at the end of the charging unit's next turn.

Options: Up to 4 Soldiers may swap their Mosin-Nagant rifle for a PPSh-41 SMG for +2 points each. A Rifle Squad may have a Soviet Medium Truck for +50 points, or an M3 halftrack for +80 points.

Tank Desant Squad – 110 points

Carried to the frontline on the hulls of tanks, these squads were small but highly flexible.

Type	Size	Move	Close Combat	Target	Save	Kill	Traits
Sergeant	1	4"	D6	4+	6+	6+	-
Soldier	1	4"	D6	4+	6+	6+	-

Unit: Sergeant (unit leader) with PPSh-41 SMG, 4 Soldiers with Mosin-Nagant rifle, Soldier with DP LMG.

Tank Assault: Infantry in this period used a variety of grenades or mines to assault armoured vehicles. All members of the unit are assumed to carry Anti-Tank Explosives. These may be placed on any model of Size 2 or greater with a Charge action. They will explode at the end of the charging unit's next turn.

Options: Up to 4 Soldiers may swap their Mosin-Nagant rifle for a PPSh-41 SMG for +2 points each. One Soldier may swap his DP LMG for a PPSh-41 SMG for -30 points.

Support Assets

The following are all the Support Assets available to a Soviet Army Infantry force.

Infantry ATR Team – 45 points

Deployed in strength, the presence of ATR teams on the battlefield was a prime concern to crews of lighter armoured vehicles.

Type	Size	Move	Close Combat	Target	Save	Kill	Traits
Lieutenant	1	4"	D6	4+	6+	6+	-
Soldier	1	4"	D6	4+	6+	6+	-

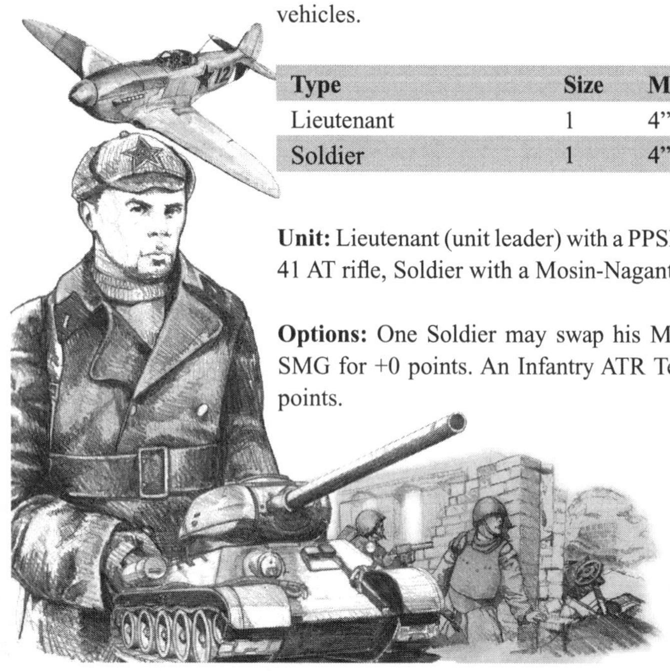

Unit: Lieutenant (unit leader) with a PPSh-41 SMG, Soldier with a PTRD-41 AT rifle, Soldier with a Mosin-Nagant rifle.

Options: One Soldier may swap his Mosin-Nagant rifle for a PPSh-41 SMG for +0 points. An Infantry ATR Team may have a Gaz-67 for +40 points.

Infantry Sniper Team – 50 points

Russian snipers became the stuff of legend in many battles, with some racking up huge tallies in their campaigns.

Type	Size	Move	Close Combat	Target	Save	Kill	Traits
Soldier	1	5”	D6	4+	6+	6+	-

Unit: Soldier (unit leader) with a Scoped Mosin-Nagant rifle, Soldier with a Mosin-Nagant rifle.

Stealthy: If a model from this unit is in Cover and did not make a Shoot action or reaction in the last turn, enemy models more than 20” away may not draw Line of Sight to it.

Ambush: This unit may be concealed in an area of terrain outside the enemy’s deployment zone before the game begins. Note the location of the unit before any models are deployed on the table. You can reveal the ambush by placing the section on the table at any time after a unit (friendly or otherwise) has completed an action. However, the unit must be revealed by the end of turn three or be counted as destroyed.

Options: An Infantry Sniper Team may have a Gaz-67 for +40 points. If it takes this option, it may not use the Ambush rule.

82mm Mortar Team – 75 points

Able to support massed human waves with accurate shelling, these mortar teams were a vital part of infantry tactics in the Soviet Army.

Type	Size	Move	Close Combat	Target	Save	Kill	Traits
Corporal	1	4”	D6	4+	6+	6+	-
Soldier	1	4”	D6	4+	6+	6+	-

Unit: Corporal (unit leader) with a Mosin-Nagant rifle, Soldier with an 82mm Mortar (FaF), 2 Soldiers with Mosin-Nagant rifles.

Options: An 82mm Mortar Team may have a Soviet Medium Truck for +50 points.

Soviet Infantry

HMG Support Team - 75 points

Though often unwieldy, the Maxim's rate of fire was sufficient to convince enemy and retreating allies alike to turn aside and seek cover.

Type	Size	Move	Close Combat	Target	Save	Kill	Traits
Corporal	1	4"	D6	4+	6+	6+	-
Soldier	1	4"	D6	4+	6+	6+	-

Unit: Corporal (unit leader) with a Mosin-Nagant rifle, Soldier with a Maxim HMG, Soldier with a Mosin-Nagant rifle.

Gun Shield: The Soldier manning the Maxim HMG counts as being in Medium Cover when shot from the Forward facing.

Options: An HMG Support Team may have a Soviet Medium Truck for +50 points.

Recon Team - 70 points

Deadly in the hands of an experienced officer, Soviet recon teams penetrated enemy lines and wreaked havoc while the main attack took place.

Type	Size	Move	Close Combat	Target	Save	Kill	Traits
Sergeant	1	4"	D6	4+	6+	6+	-
Soldier	1	4"	D6	4+	6+	6+	-

Unit: Sergeant (unit leader) with a PPSh-41 SMG, 3 Soldiers with PPSh-41 SMGs.

Stealthy: If a model from this unit is in Cover and did not make a Shoot action or reaction in the last turn, enemy models more than 20" away may not draw Line of Sight to it.

Ambush: This unit may be concealed in an area of terrain outside the enemy's deployment zone before the game begins. Note the location of the unit before any models are deployed on the table. You can reveal the ambush by placing the section on the table at any time after a unit (friendly or otherwise) has completed an action. However, the unit must be revealed by the end of turn three or be counted as destroyed.

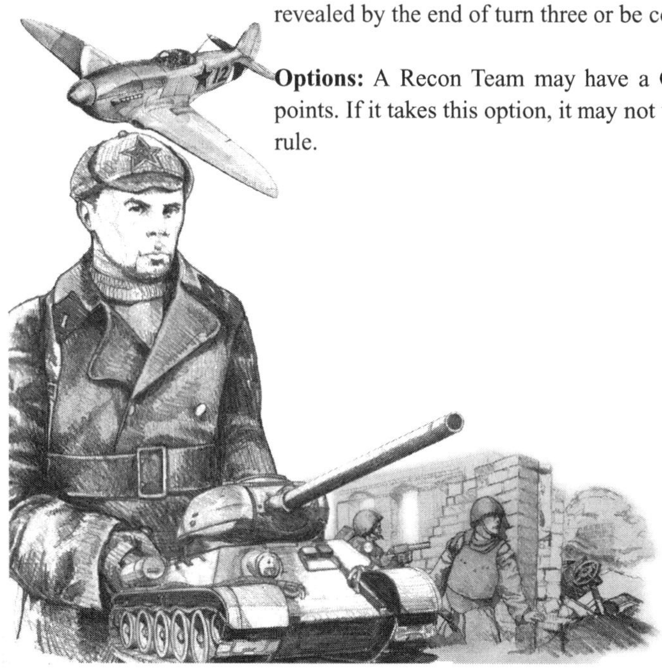

Options: A Recon Team may have a Gaz-67 for +40 points. If it takes this option, it may not use the Ambush rule.

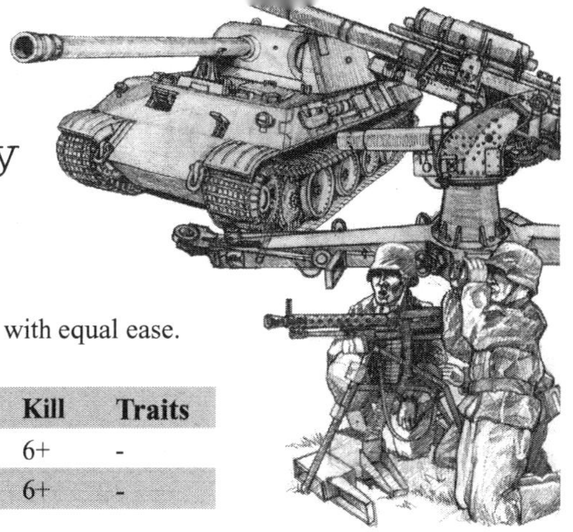

76mm Field Gun - 95 points

A solid field gun, the 76mm shells fired were capable of blasting apart tank and infantry attacks with equal ease.

Type	Size	Move	Close Combat	Target	Save	Kill	Traits
Sergeant	1	4"	D6	4+	6+	6+	-
Soldier	1	4"	D6	4+	6+	6+	-

Unit: Sergeant (unit leader) with a PPSh-41 SMG, 1 Soldier with a 76mm Field Gun (FaF), 2 Soldiers with Mosin-Nagant rifles.

Crew Weapon: The 76mm Field Gun can only shoot if at least one crew member is in base contact with it.

Cumbersome: A Ready action must be taken before the 76mm Field Gun may be moved. The Move score of all models in the unit depends on the number of models still manning the weapon.

4 crew = 4" maximum move,
3 crew = 3" maximum move,
2 crew = 2" maximum move and
1 crew = No movement possible.

If the crew choose to abandon the 76mm Infantry Gun, they may move normally.

Gun Shield: The crew count as being in Medium Cover when shot from the Forward facing.

Assault Engineer Squad - 190 points

Soviet infantry were famed for their tough and dogged approach to battle. Assault Engineers were the most stubborn of them. It was their duty to do the most gruelling city fighting, engineering work under enemy fire, take fortified buildings or clean mine fields.

Type	Size	Move	Close Combat	Target	Save	Kill	Traits
Senior Engineer	1	5"	D6+1	4+	6+	6+	-
Engineer	1	5"	D6+1	4+	6+	6+	-
Soldier	1	5"	D6+1	4+	6+	6+	-

Team 1: Senior Engineer (unit leader) with a PPSh-41 SMG, 4 Soldiers with Mosin-Nagant rifles.

Team 2: Engineer with PPSh-41 SMG, 4 Soldiers with Mosin-Nagant rifle.

Ambush: This unit may be concealed in an area of terrain outside the enemy's deployment zone before the game begins. Note the location of the unit before any models are deployed on the table. You can reveal the ambush by placing the section on the table at any time after a unit (friendly or otherwise) has completed an action. However, the unit must be revealed by the end of turn three or be counted as destroyed.

Elite Training: The unit gains a +1 bonus to their Armour Saves when in Cover. They may also re-roll their Close Combat dice.

Soviet Infantry

Dedicated: If Suppressed, roll a D6. If you can roll equal to or less than the number of soldiers left in the unit, you can ignore the effects of the Suppression.

Teams: The Engineer may act as unit leader to Team 2, splitting off to act as a separate unit.

Tank Assault: Infantry in this period used a variety of grenades or mines to assault armoured vehicles. All members of the unit are assumed to carry Anti-Tank Explosives. These may be placed on any model of Size 2 or greater with a Charge action if the target ends up within the Lethal Range of the charging unit. They will explode at the end of the charging unit's next turn.

Options: Drop Team 2 for –90 points. Up to 1 Soldier of either team may be equipped with a Bazooka in addition to his rifle for +20 points each. Up to 1 Soldier of either team may swap his rifle for a DP-LMG for +30 points each. Up to 4 Soldiers of either team may swap their rifles for a PPSh-41 SMG for +2 points each. Up to 1 Soldier of either team may swap his rifle for a Flamethrower for +10 points each. Either Team may be equipped with Improved Anti-Tank Explosives for +30 points per section. An Assault Engineer Squad may have a Soviet Medium Truck for +50 points or an M3 halftrack for +80 points. If a vehicle is taken, the Squad may no longer use the Ambush rule.

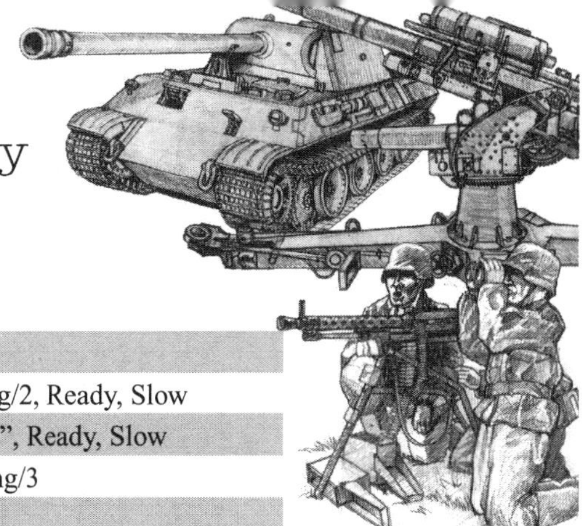

Late War Soviet Armoury

Weapon	Range	Damage	Trait
76mm Field Gun	36”	D10+2	Lethal Zone/1”, Piercing/2, Ready, Slow
82mm Mortar	48”	D6	Artillery, Lethal Zone/3”, Ready, Slow
Anti-Tank Explosives	Placed	D10	Lethal Zone /2”, Piercing/3
Bazooka	18”	D10+2	Lethal Zone/1”, Slow
DP LMG	24”	2xD6	Auto, Prone
Flamethrower	10”	3xD6	Flame, Slow
Improved Anti-Tank Explosives	Placed	2xD10+2	Lethal Zone /2”, Piercing/3
Maxim HMG	48”	2xD6+1	Auto, Prone
Mosin-Nagant rifle	24”	D6	-
PPSh-41 SMG	12”	3xD6	-
PTRD-41 AT rifle	36”	D6+2	Slow
Scoped Mosin-Nagant	36”	D6	Accurate

76mm Field Gun: One of the best artillery pieces of World War II, this weapon was able to punch through German armour with ease.

82mm Mortar: Though vulnerable to assault, mortar positions were quick to set up in battle, and provided a steady rain of fire on an increasingly dispirited enemy

Anti-Tank Explosives: A combination of high-powered grenades and make-shift charges, these explosives give infantry a fighting chance against armoured vehicles, though their use is only advised in built-up areas.

Bazooka: The primary anti-tank weapon of American armed forces, many found their way into other Allied forces under the Lend-Lease agreement.

DP LMG: Cheap and easy to mass-produce, the DP LMG could withstand a great amount of mistreatment and still remain able to fire effectively.

Flamethrower: As dangerous to the user as the enemy when under fire, the flamethrower was very effective at clearing infantry out of cover, and soon achieved a fearsome reputation.

Improved Anti-Tank Explosives: Dedicated satchel charges were used by elite units in order to immobilise and destroy enemy armour, though such attacks were extremely dangerous.

Maxim HMG: Mounted on a wheeled trolley, the Maxim HMG had a distinctive chatter, and was suitable for mowing down German and Russian infantry alike.

Mosin-Nagant rifle: The standard rifle of the Soviet Army, this weapon was manufactured in its millions.

PPSh-41 SMG: With either box or drum magazines, the PPSh-41 was an effective and low maintenance submachine gun.

PTRD-41 AT rifle: Prior to the arrival of the Bazooka, this was the only anti-tank weapon the Soviet army possessed, and soldiers had to target specific weak points on enemy armour.

Scoped Mosin-Nagant: The weapon of Vasily Zaitsev, this version of the Mosin-Nagent was able to stand up to the most brutal of environments and remain deadly accurate.

LATE WAR SOVIET VEHICLES AND AIR ASSETS

The following Assets may be used by any Late War Soviet Army force.

Support Assets

The following are all the Support Assets available to a Soviet Army force.

BA-64 scout car - 70 points

Despite its light armament, the BA-64 (nicknamed the Bobik) had a good turn of speed and was superior in the reconnaissance role.

Type	Size	Move	Close Combat	Facing	Target	Save	Kill	Traits
BA-64	2	8"	2xD10	Front	7+	5+	8+	Hits/2, Lumbering, Wheels/16"
				Sides	6+	6+	8+	
				Rear	6+	6+	7+	

Unit: BA-64 with an MG.

GAZ-67 - 40 points

Designed as a staff vehicle, the Gaz-67 was a rugged all-terrain vehicle, similar to the American Jeep.

Type	Size	Move	Close Combat	Target	Save	Kill	Traits
GAZ-67	2	8"	D10	6+	6+	7+	Hits/2, Lumbering, Wheels/16"

Unit: GAZ-67

Transport: The GAZ-67can carry up to 5 Size points of models (each maximum Size 1).

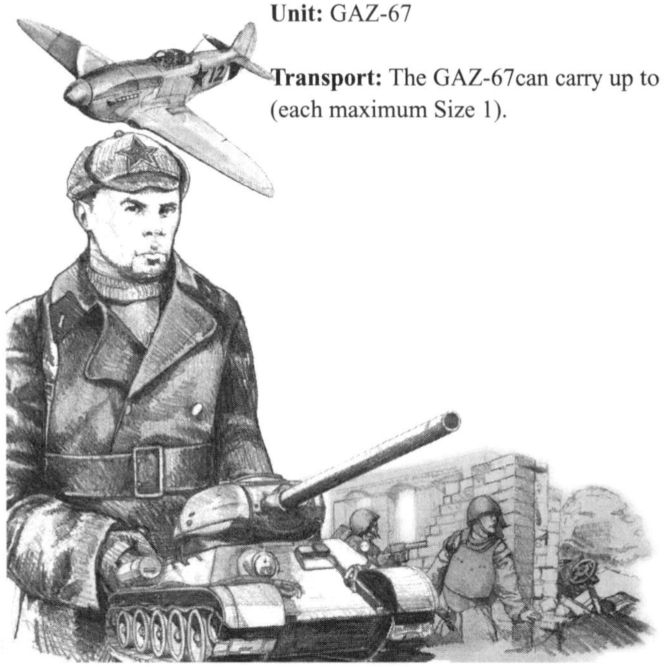

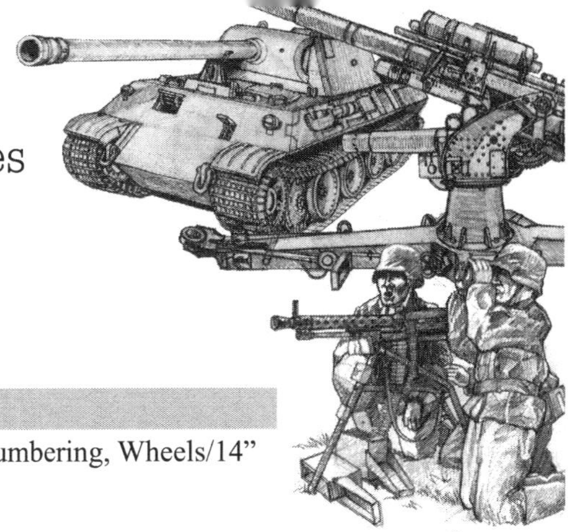

GAZAAA-37 Truck - 90 points

This towed anti-aircraft weapon provided much-needed protection to exposed ground troops.

Type	Size	Move	Close Combat	Target	Save	Kill	Traits
GAZAAA-37 Truck	3	7”	D10	6+	6+	8+	Hits/2, Lumbering, Wheels/14”

Unit: GAZAAA-37 Truck with a 37mm L74 AA Gun (FaF).

M3 Halftrack - 80 points

An American-built armoured vehicle, the M3 was used by the Soviet Army under the Lend-Lease agreement, and had dozens of different variants.

Type	Size	Move	Close Combat	Facing	Target	Save	Kill	Traits
M3 Halftrack	3	7”	2xD10	Front	7+	5+	8+	Hits/2, Lumbering, Tracks/14”
				Sides	6+	6+	7+	
				Rear	6+	6+	7+	

Unit: M3 with a Pintle MG (FaF).

Transport: The M3 can carry up to 12 Size points of models (each maximum Size 1).

Options: A Pintle MG (FaA) may be added for +35 points.

Soviet Medium Truck - 50 points

A variety of designs were prevalent in the Soviet Army, all trucks used to carry men and equipment across the vast Russian homeland and beyond.

Type	Size	Move	Close Combat	Target	Save	Kill	Traits
Medium Truck	4	6”	D10	6+	6+	7+	Hits/3, Lumbering, Wheels/12”

Unit: Soviet Medium Truck.

Transport: The Medium Truck can carry up to 10 Size points of models (each maximum Size 1).

Armour Assets

The following are all the Armour Assets available to a Soviet Army force.

Iosif Stalin 2 Tank - 490 points

Named after the leader of the Soviet Union, the Iosif Stalin 2 has thick armour to defeat the best of German guns, while its own armament was capable of destroying any tank it faced.

Type	Size	Move	Close Combat	Facing	Target	Save	Kill	Traits
Iosif Stalin 2	4	4”	4xD10	Front	9+	3+	13+	Hits/3, Lumbering, Multifire/2, Tracks/7”
				Sides	9+	3+	13+	
				Rear	8+	4+	12+	

Unit: Iosif Stalin 2 with a 122mm D25 anti-tank gun, a Pintle MG and an MG.

Co-Axial MG: To improve target acquisition, MGs on vehicles were often mounted co-axial to the main gun. The co-axial MG must always use the same Fire Zone as the main gun. As soon as a Fire Zone is placed within range of the co-axial MG, you may add +1 to all Damage Dice of the main gun.

Tank Rider: The Iosif Stalin 2 can transport up to 6 models (maximum Size 1) on its engine deck. The transported infantry models count as in medium Cover against shooting attacks from the Forward facing, but do not receive Cover from other facings.

KV-1C Heavy Tank - 390 points

The KV-1C was an upgraded version of the heavy tank that had caused real problems for the Gemans in the early stages of the war. Its mobility and armament were inferior, however, and was outlassed by the later German tanks.

Type	Size	Move	Close Combat	Facing	Target	Save	Kill	Traits
KV-1 C	4	3”	4xD10	Front	9+	3+	13+	Hits/3, Lumbering, Multifire/2, Tracks/6”
				Sides	9+	4+	12+	
				Rear	8+	4+	12+	

Unit: KV-1 C tank with a 76.2mm F34 tank gun, an MG and an MG (FaF).

Co-Axial MG: To improve target acquisition, MGs on vehicles were often mounted co-axial to the main gun. One MG has a co-axial mount. The co-axial MG must always use the same Fire Zone as the main gun. As soon as a Fire Zone is placed within range of the co-axial MG, you may add +1 to all Damage Dice of the main gun.

Tank Rider: The KV-1 C can transport up to 6 models (maximum Size 1) on its engine deck. The transported infantry models count as in medium Cover against shooting attacks from the Forward facing, but do not receive Cover from other facings.

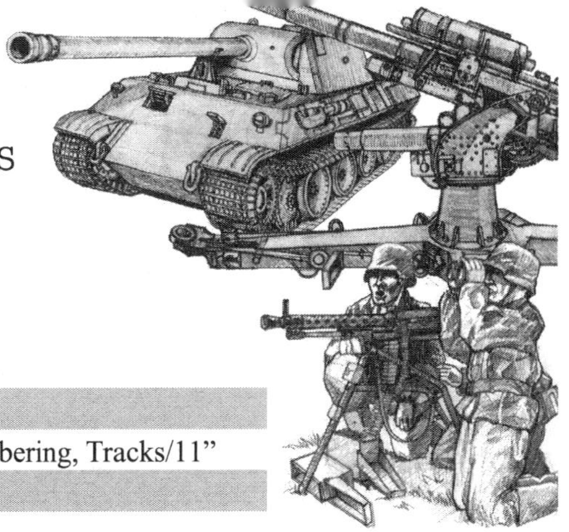

SU-85 Tankhunter - 185 points

Based on the chassis of a T-34, this was the first dedicated tank destroyer of the Soviet Army.

Type	Size	Move	Close Combat	Facing	Target	Save	Kill	Traits
SU-85	4	6"	4xD10	Front	8+	3+	12+	Hits/2, Lumbering, Tracks/11"
				Sides	7+	5+	11+	
				Rear	7+	5+	10+	

Unit: SU-85 tank with an 85mm L54 anti-tank gun (FaF).

SU-100 Tankhunter - 210 points

Developed late in the war, Russia was able to produce large numbers of this tank destroyer, and many were in service during 1945.

Type	Size	Move	Close Combat	Facing	Target	Save	Kill	Traits
SU-100	4	5"	4xD10	Front	8+	3+	12+	Hits/2, Lumbering, Tracks/10"
				Sides	7+	5+	11+	
				Rear	7+	5+	10+	

Unit: SU-100 tank with a 100mm L56 anti-tank gun (FaF).

T-34/43 Tank - 190 points

Constantly refined during the war, the T-34 was a definitive tank in many ways and was responsible for influencing future designs from all nations.

Type	Size	Move	Close Combat	Facing	Target	Save	Kill	Traits
T-34/43	4	6"	4xD10	Front	8+	4+	12+	Hits/2, Lumbering, Multifire/2, Tracks/11"
				Sides	7+	5+	11+	
				Rear	7+	5+	10+	

Unit: T-34/43 tank with a 76.2mm F34 tank gun, an MG and an MG (FaF).

Co-Axial MG: To improve target acquisition, MGs on vehicles were often mounted co-axial to the main gun. One MG has a co-axial mount. The co-axial MG must always use the same Fire Zone as the main gun. As soon as a Fire Zone is placed within range of the co-axial MG, you may add +1 to all Damage Dice of the main gun.

Tank Rider: The T-34/43 can transport up to 6 models (maximum Size 1) on its engine deck. The transported infantry models count as in medium Cover against shooting attacks from the Forward facing, but do not receive Cover from other facings.

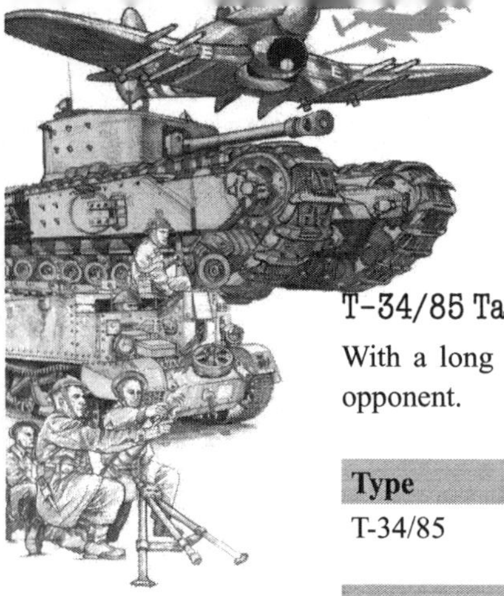

Soviet Vehicles

T-34/85 Tank - 240 points

With a long 85mm main gun, the T-34/85 was a substantial improvement on the original design, and a fearsome opponent.

Type	Size	Move	Close Combat	Facing	Target	Save	Kill	Traits
T-34/85	4	5”	4xD10	Front	8+	3+	12+	Hits/2, Lumbering, Multifire/2, Tracks/10”
				Sides	7+	4+	11+	
				Rear	7+	5+	10+	

Unit: T-34/85 tank with an 85mm L54 tank gun, an MG and an MG (FaF).

Co-Axial MG: To improve target acquisition, MGs on vehicles were often mounted co-axial to the main gun. One MG has a co-axial mount. The co-axial MG must always use the same Fire Zone as the main gun. As soon as a Fire Zone is placed within range of the co-axial MG, you may add +1 to all Damage Dice of the main gun.

Tank Rider: The T-34/85 can transport up to 6 models (maximum Size 1) on its engine deck. The transported infantry models count as in medium Cover against shooting attacks from the Forward facing, but do not receive Cover from other facings.

Air Assets

The following are all the Air Assets available to a Soviet Army force.

Ilyushin Sturmovik II-2 - 240 points

Coining the term ‘flying tank’, the Il-2 was extremely well-armoured for an attack aircraft and proved deadly against ground targets.

Type	Size	Move	Close Combat	Target	Save	Kill	Traits
Ilyushin Sturmovik II-2	5	Loiter Cruise	4xD10	6+	4+/6+	8+	Hits/3, Multifire/4

Unit: Ilyushin Sturmovik II-2 with two Vya-23 cannon (FaF), two ShKAS LMG (FaF), and an MG (FaR).

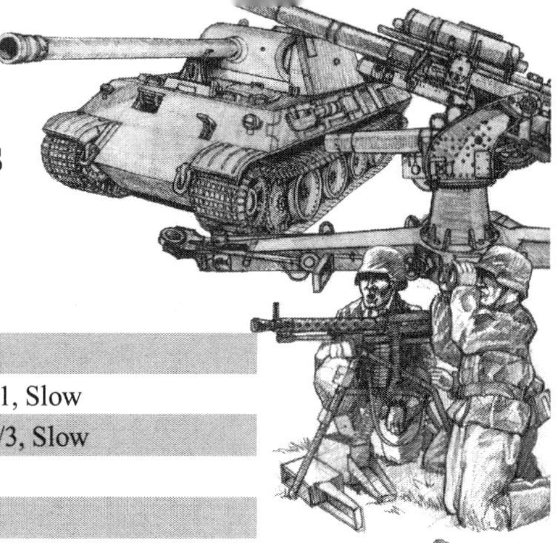

Soviet Vehicle Armoury

Weapon	Range	Damage	Trait
100mm D-10 anti-tank gun	60"	D10+4	Lethal Zone/1",Multihit, Piercing/1, Slow
122mm D-25 anti-tank gun	60"	D10+4	Lethal Zone/1", Multihit, Piercing/3, Slow
37mm L74 AA gun	36"	4xD6+2	AA, Piercing/1, Slow
76.2mm F34 tank gun	48"	D10+2	Lethal Zone/1", Slow
85mm L54 anti-tank gun	60"	D10+3	Lethal Zone/1", Piercing/2, Slow
MG	24"	3xD6	Auto
Pintle MG	24"	3xD6	AA, Auto
ShKAS LMG	24"	2xD6	
Vya-23 cannon	24"	2xD6+2	Piercing/1

100mm D-10 anti-tank gun: Capable of firing very heavy shells, this anti-tank gun could blast apart any German armour it faced.

122mm D-25 anti-tank gun: In tests, this mighty gun could destroy a German Tiger at massive distances, and was an excellent choice for the Iosif Stalin tanks.

37mm L74 AA gun: Providing valuable cover for advancing forces, the presence of this gun on the battlefield forced enemy air pilots not to linger over their targets.

76.2mm F34 tank gun: A versatile weapon, this gun was mounted on several armoured vehicles of the Soviet Army, though it was superceded by more recent designs in the late war period.

85mm L54 anti-tank gun: First appearing on the Su-85, this gun was quickly adopted for the T-34, where it fast out-paced the performance of the F34.

MG: Covering a wide variety of weapons, machine guns mounted on vehicles are used to suppress enemy infantry and provide supporting fire.

Pintle MG: When seated on a pintle mount, a machine guns gains a greater field of fire and elevation, allowing it to be used to engage low-flying aircraft.

ShKAS LMG: A 7.62mm machine gun, this weapon found use on many Russian fighters and bombers.

Vya-23 cannon: Intended to be used in the anti-tank role, pilots usually had to attack their targets from the rear or side to gain a noticeable effect.

LATE WAR US ARMY INFANTRY

Having adopted an isolationist policy early in the war, when America entered the conflict it did so wholeheartedly. The presence of American personnel and equipment made the liberation of Europe possible, and US troops waged a hard fought war across Italy, France, Belgium, Holland and into Germany. Well-equipped and with huge resources of manpower, it was the US that tipped the balance on the western front.

Platoon Structure

The standard squad size for most US Army platoons is twelve soldiers, consisting of two teams. The squad is led by a Sergeant, while the second team is led by a Corporal.

Three such sections make up a standard platoon, led by a Lieutenant, who is in turn assisted by a Sergeant. Three platoons make a rifle (infantry) company.

A platoon that has been on operational duty for some time may vary wildly from this, however. Heavy losses can result in fewer and smaller sections, while a platoon involved in heavy combat may become 'reinforced' with more soldiers, giving it more sections.

You can make any platoon with the US Army list, from one that has just been deployed, to a grizzled set of veterans that have taken heavy losses or been reinforced.

Each unit in a platoon is considered to be an Asset of some type. You may include the following Assets in your platoon, so long as you obey the minimum and maximum units listed for each. You may take one platoon for every 1,500 points in your force.

0-1 Command Assets
2-4 Squad Assets
0-1 Support Asset per Squad Asset
0-3 Armour Assets
0-3 Air Assets

For details on more Support, Armour and Air Assets see the US Vehicle and Air Assets List on page 134).

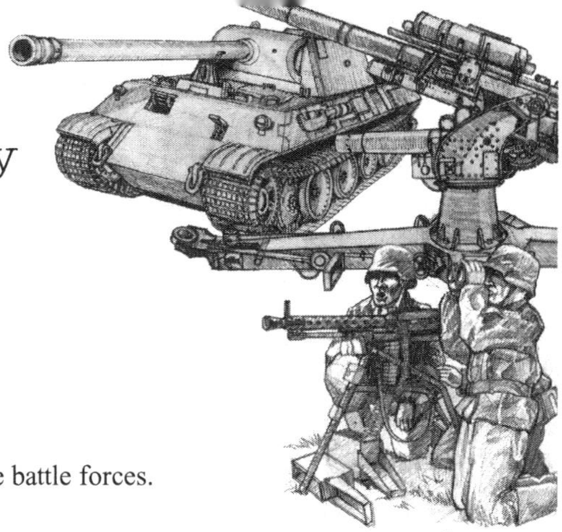

Command Assets

The following are all the Command Assets available to a US Army Infantry force.

Command Squad – 90 points

With powerful squads and support units at his disposal, the Lieutenant and his squad led capable battle forces.

Type	Size	Move	Close Combat	Target	Save	Kill	Traits
Lieutenant	1	4”	D6	4+	6+	6+	-
Staff Sergeant	1	4”	D6	4+	6+	6+	-
Soldier	1	4”	D6	4+	6+	6+	-

Team 1: Lieutenant (unit leader) with a M3 SMG, Staff Sergeant with a M3 SMG, Soldier with a M1 Garand rifle.

Team 2: Soldier with a Bazooka, Soldier with a M1 Garand rifle.

Teams: One Soldier may act as unit leader to Team 2, splitting off to act as a separate unit.

Supporting Artillery: This unit may purchase up to 3 uses of Supporting Artillery. Calling in Supporting Artillery requires a Ready action. This is used as an Artillery attack, but the initial Impact Point must be in the unit’s Line of Sight. There is no minimum range for this attack (yes, you can call it down right on your own head!). Light Artillery costs +50 points per use, while Heavy Artillery costs +100 points per use.

Options: Any member of the squad may swap their M3 SMG or M1 Garand for an M1 Carbine for +2 points each. Drop Team 2 for –55 points. Add up to one additional Team 2 for +55 points.

Squad Assets

The following are all the Squad Assets available to a US Army Infantry force.

Rifle Squad – 180 pts

Enjoying the best logistical support possible, the US infantryman came to expect victory.

Type	Size	Move	Close Combat	Target	Save	Kill	Traits
Sergeant	1	4”	D6	4+	6+	6+	-
Corporal	1	4”	D6	4+	6+	6+	-
Soldier	1	4”	D6	4+	6+	6+	-

Team 1: Sergeant (unit leader) with a M3 SMG, 7 Soldiers with M1 Garand rifles.

Team 2: Corporal with a M3 SMG, a Soldier with M1918 BAR, 2 Soldiers with M1 Garand rifles.

Teams: The Corporal may act as unit leader to Team 2, splitting off to act as a separate unit.

Tank Assault: Infantry in this period used a variety of grenades or mines to assault armoured vehicles. All members of

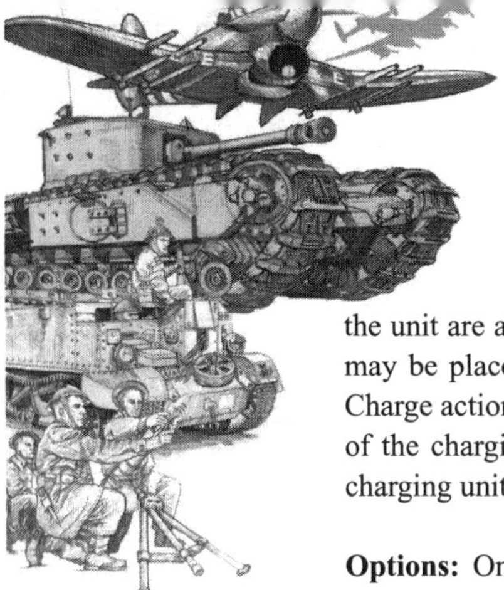

the unit are assumed to carry Anti-Tank Explosives. These may be placed on any model of Size 2 or greater with a Charge action if the target ends up within the Lethal Range of the charging unit. They will explode at the end of the charging unit's next turn.

Options: One Soldier may be equipped with a Grenade Launcher fitting for his M1 Garand for +10 points. One Soldier per team may swap his M1 Garand for an M1 Carbine for +2 points. Drop Team 2 for -80 points.

Support Assets

The following are all the Support Assets available to a US Army Infantry force.

HMG Company Support Team – 80 points

Supporting the platoon's advance HMG teams suppressed enemies in cover and annihilated any who tried to counterattack.

Type	Size	Move	Close Combat	Target	Save	Kill	Traits
Corporal	1	4”	D6	4+	6+	6+	-
Soldier	1	4”	D6	4+	6+	6+	-

Unit: Corporal (unit leader) with a M3 SMG, Soldier with a M2 .50 Cal HMG, Soldier with a M1 Garand rifle

Unit Options: Any member of the team may swap their M3 SMG or M1 Garand for an M1 Carbine for +2 points each.

Bazooka Platoon Support Team – 50 points

The famous Bazooka allowed nimble infantry teams to engage tanks with a good degree of success.

Type	Size	Move	Close Combat	Target	Save	Kill	Traits
Soldier	1	4”	D6	4+	6+	6+	-

Unit: Soldier (unit leader) with a Bazooka and a M3 SMG, Soldier with a M1 Garand rifle.

Options: The M1 Garand may be swapped for an M1 Carbine for +2 points.

81mm Mortar Company Support Team – 70 points

Laying down successive bombardments, mortar teams forced enemies to redeploy units in order to escape the worst effects of this light artillery.

Type	Size	Move	Close Combat	Target	Save	Kill	Traits
Corporal	1	4"	D6	4+	6+	6+	-
Soldier	1	4"	D6	4+	6+	6+	-

Unit: Corporal (unit leader) with a M3 SMG, Soldier with an 81mm Mortar (FaF), Soldier with a M1 Garand rifle.

Options: Any member of the team may swap their M3 SMG or M1 Garand for an M1 Carbine for +2 points each.

Sniper Platoon Support Team – 50 points

Spread across the battlefield, multiple sniper teams could play havoc with an enemy's plans.

Type	Size	Move	Close Combat	Target	Save	Kill	Traits
Soldier	1	5"	D6	4+	6+	6+	-

Unit: Soldier (unit leader) with a Scoped M1903A Springfield, Soldier with a M1 Garand rifle.

Stealthy: If a model from this unit is in Cover and did not make a Shoot action or reaction in the last turn, enemy models more than 20" away may not draw Line of Sight to it.

Ambush: This unit may be concealed in an area of terrain outside the enemy's deployment zone before the game begins. Note the location of the unit before any models are deployed on the table. You can reveal the ambush by placing the section on the table at any time after a unit (friendly or otherwise) has completed an action. However, the unit must be revealed by the end of turn three or be counted as destroyed.

57mm AT Gun – 85 points

Capable of giving German armour pause for thought, anti-tank guns were vital in supporting their infantry.

Type	Size	Move	Close Combat	Target	Save	Kill	Traits
Corporal	1	4"	D6	4+	6+	6+	-
Soldier	1	4"	D6	4+	6+	6+	-

Unit: Corporal (unit leader) with a M3 SMG, Soldier with a 57mm anti-tank gun (FaF), 2 Soldiers with M1 Garand rifles.

Crew Weapon: The 57mm anti-tank gun can only shoot if at least one crew member is in base contact with it.

Cumbersome: A Ready action must be taken before the 57mm anti-tank gun may be moved. The Move score of all models in the unit depends on the number of models still manning the weapon.

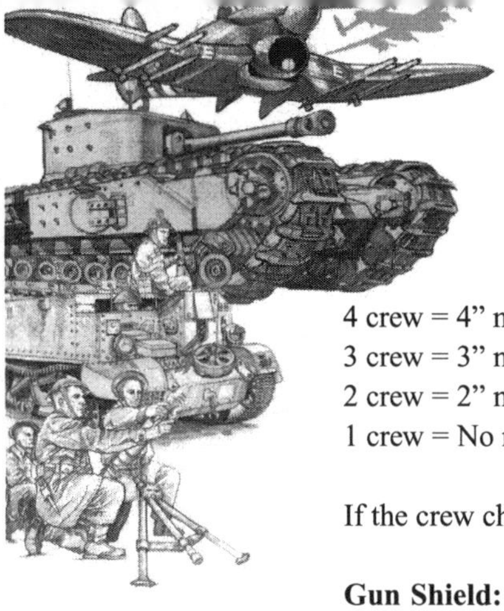

4 crew = 4” maximum move,
3 crew = 3” maximum move,
2 crew = 2” maximum move and
1 crew = No movement possible.

If the crew choose to abandon the 57mm anti-tank gun, they may move normally.

Gun Shield: The Corporal and Soldiers count as being in Medium Cover when shot from the Forward facing.

Options: Any member of the team may swap their M3 SMG or M1 Garand for an M1 Carbine for +2 points each.

Combat Engineer Squad – 255 points

US combat Engineers served in every major US operation and their job was often the most dangerous. It was their duty to do engineering work under enemy fire, take fortified buildings or clear mine fields.

Type	Size	Move	Close Combat	Target	Save	Kill	Traits
Sergeant	1	5”	D6+1	4+	6+	6+	-
Soldier	1	5”	D6+1	4+	6+	6+	-

Engineer Team: Sergeant (unit leader) with a M1 Carbine, 3 Soldiers with M1 Carbines.

BAR Team: Gunner with a M1918 BAR, Soldier with a M1 Garand rifle.

Bazooka Team: Gunner with a Bazooka and M1 Carbine, a Soldier with M1 Garand rifle.

Flamethrower Team: Gunner with a M2-2 Flamethrower, Soldier with a M1 Garand rifle.

Ambush: This unit may be concealed in an area of terrain outside the enemy’s deployment zone before the game begins. Note the location of the unit before any models are deployed on the table. You can reveal the ambush by placing the section on the table at any time after a unit (friendly or otherwise) has completed an action. However, the unit must be revealed by the end of turn three or be counted as destroyed.

Elite Training: The unit gains a +1 bonus to their Armour Saves when in Cover. They may also re-roll their Close Combat dice.

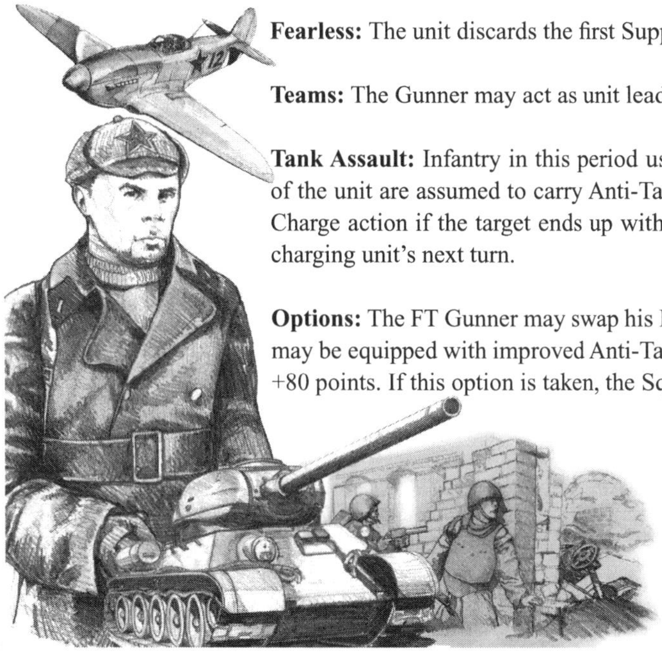

Fearless: The unit discards the first Suppression dice it suffers every turn.

Teams: The Gunner may act as unit leader to their Teams, splitting off to act as a separate unit.

Tank Assault: Infantry in this period used a variety of grenades or mines to assault armoured vehicles. All members of the unit are assumed to carry Anti-Tank Explosives. These may be placed on any model of Size 2 or greater with a Charge action if the target ends up within the Lethal Range of the charging unit. They will explode at the end of the charging unit’s next turn.

Options: The FT Gunner may swap his M2-2 Flamethrower for an M1919A6 .30 Cal HMG +40 points. The whole unit may be equipped with improved Anti-Tank Explosives for +30 points. An Engineer Squad may have 1 M3 halftrack for +80 points. If this option is taken, the Squad no longer benefits from the Ambush rule.

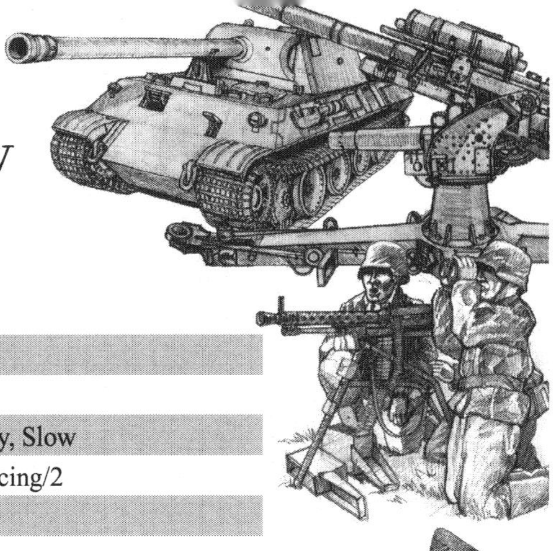

Late War US Armoury

Weapon	Range	Damage	Traits
57mm anti-tank gun	36"	D10+2	Lethal Zone/1", Ready, Slow
81mm Mortar	48"	D6	Artillery, Lethal Zone/3", Ready, Slow
Anti-Tank Explosives	Placed	D10	Lethal Zone/2", One-Shot, Piercing/2
Bazooka	18"	D10+2	Lethal Zone/ 1", Slow
Grenade Launcher	24"	D6	Lethal Zone/ 2", Piercing/1
Heavy Artillery	Special	2xD6+1	Lethal Zone/4", One-Shot
Improved Anti-Tank Explosives	Placed	2xD10+2	Lethal Zone/2", One-Shot, Piercing/3
Light Artillery	Special	D6	Lethal Zone/4", One-Shot
M1 Carbine	18"	2xD6	-
M1 Garand Rifle	24"	D6	Auto
M1918 BAR	24"	3xD6	Auto
M2 .50 Cal HMG	48"	2xD6+1	Auto, Piercing/1, Prone
M2-2 Flamethrower	10"	3xD6	Flame, Slow
M3 SMG	12"	3xD6	-
Scoped M1903A Springfield	36"	D6	Accurate

57mm anti-tank gun: Though outpaced towards the end of the war by new designs in German armour, few tank crews would willingly leave themselves open to sustained attack by a battery of these guns.

81mm Mortar: Though vulnerable to assault, mortar positions were quick to set up in battle, and provided a steady rain of fire on an increasingly dispirited enemy.

Anti-Tank Explosives: A combination of high-powered grenades and make-shift charges, these explosives give infantry a fighting chance against armoured vehicles, though their use is only advised in built-up areas.

Bazooka: With a shaped-charged warhead, the Bazooka finally gave infantry the means to destroy enemy tanks, and the design was to be copied by every major military force.

Grenade Launcher: Clipping onto the barrel of the M1 Garand rifle, this accessory allowed a soldier to launch grenades a good distance, providing an ad hoc mortar like role.

Heavy Artillery: At the call of platoon leaders, heavy artillery could literally blast infantry out of cover, and was effective against all but the heaviest of armoured vehicles.

Improved Anti-Tank Explosives: Dedicated satchel charges were used by elite units in order to immobilise and destroy enemy armour, though such attacks were extremely dangerous.

Light Artillery: Using smaller and short-ranged guns, light artillery is primarily an anti-infantry weapon, forcing enemy troops into cover and disrupting their attacks as an assault is launched.

US Infantry

M1 Carbine: Based on the Garand, the M1 Carbine could be used in much the same way as a modern assault rifle.

M1 Garand Rifle: Replacing the Springfield as the US Army's main firearm, the Garand was an effective self-loading rifle.

M1918 BAR: A light machine gun, the BAR had selectable rates of fire, though was initially hampered by its small magazine.

M2 .50 Cal HMG: With integrated bipod, the Browning was an effective, though cumbersome, weapon.

M2-2 Flamethrower: Though to be made obsolete by advances in flame throwing tanks, the M2 was a vicious weapon in close quarters, and had significant advantages in built up areas.

M3 SMG: Nicknamed the grease gun, the M3 replaced the Thompson and was a good short-ranged weapon.

Scoped M1903A Springfield: Though replaced by the Garand, the Springfield found plenty of life in the late war period as a scoped sniper rifle.

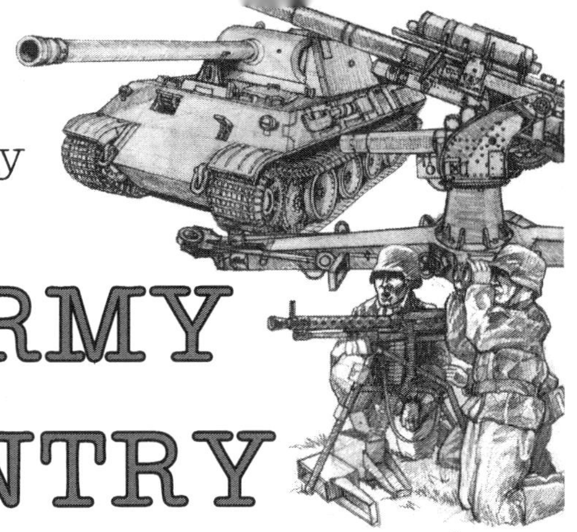

LATE WAR US ARMY ARMOURED INFANTRY

These completely motorised platoons were mounted in Halftracks and other vehicles. Their tactical role was to provide support for the fast-moving armoured divisions. Able to deploy and counterattack with lightning speed, an armoured infantry platoon could assault an enemy position, occupy it and then move on to a new target before enemy reinforcements could gather enough strength to attack. When combined with armour these platoons could be all but unstoppable.

Platoon Structure

The standard squad size for most US Army platoons is twelve soldiers, consisting of two teams. The squad is led by a Sergeant, while the second team is led by a Corporal.

Three such sections make up a standard platoon, led by a Lieutenant who is in turn assisted by a Sergeant. Three platoons make a an armoured infantry company.

A platoon that has been on operational duty for some time may vary wildly from this, however. Heavy losses can result in fewer and smaller sections, while a platoon involved in heavy combat may become 'reinforced' with more soldiers, giving it more sections.

You can make any platoon with the US Army list, from one that has just been deployed, to a grizzled set of veterans that have taken heavy losses or been reinforced.

Each unit in a platoon is considered to be an Asset of some type. You may include the following Assets in your platoon, so long as you obey the minimum and maximum units listed for each. You may take one platoon for every 1,500 points in your force.

0-1 Command Assets
2-4 Squad Assets
0-1 Support Asset per Squad Asset
0-1 Armour Assets Asset per Squad Asset
0-3 Air Assets

For details on more Support, Armour and Air Assets see the US Vehicle and Air Assets List on page 134).

Command Assets

The following are all the Command Assets available to a US Army Armoured Infantry force.

Command Squad - 150 points

With powerful squads and support units at his disposal, the Lieutenant and his squad led capable battle forces.

Type	Size	Move	Close Combat	Target	Save	Kill	Traits
Lieutenant	1	4”	D6	4+	6+	6+	—
Staff Sergeant	1	4”	D6	4+	6+	6+	—
Soldier	1	4”	D6	4+	6+	6+	—

Unit: Lieutenant (unit leader) with a M3 SMG, Staff Sergeant with a M3 SMG, 9 Soldiers with M1 Garand rifles.

Supporting Artillery: This unit may purchase up to 3 uses of Supporting Artillery. Calling in Supporting Artillery requires a Ready action. This is used as an Artillery attack, but the initial Impact Point must be in the unit's Line of Sight. There is no minimum range for this attack (yes, you can call it down right on your own head!). Light Artillery costs +50 points per use, while Heavy Artillery costs +100 points per use.

Options: Any member of the Command squad may swap their M3 SMG or M1 Garand for an M1 Carbine for +2 points each. One Soldier may be equipped with a BAR for +20 points. The Squad may have an M3 halftrack for +80 points.

Squad Assets

The following are all the Squad Assets available to a US Army Armoured Infantry force.

Rifle Squad - 140 pts

Enjoying the best logistical support, the US infantry came to expect victory.

Type	Size	Move	Close Combat	Target	Save	Kill	Traits
Sergeant	1	4”	D6	4+	6+	6+	—
Corporal	1	4”	D6	4+	6+	6+	—
Soldier	1	4”	D6	4+	6+	6+	—

Team 1: Sergeant (unit leader) with a M3 SMG, 4 Soldiers with M1 Garand rifles.

Team 2: Corporal with a M3 SMG, 4 Soldiers with M1 Garand rifles.

Teams: The Corporal may act as unit leader to Team 2, splitting off to act as a separate unit.

Tank Assault: Infantry in this period used a variety of grenades or mines to assault armoured vehicles. All members of the unit are assumed to carry Anti-Tank Explosives. These may be placed on any model of Size 2 or greater with a Charge action if the target ends up within the Lethal Range of the charging unit. They will explode at the end of the charging unit's next turn.

Options: The Sergeant and Corporal may swap their M3 SMG for an M1 Carbine for +2 points each. One Soldier per team may be equipped with a Grenade Launcher fitting for his M1 Garand for +10 points. One Soldier per team may be equipped with a BAR for +20 points. One Soldier per team may swap their M1 Garand for an M1 Carbine for +2 points each. One Soldier may be equipped with a Bazooka for +20 points. Drop Team 2 for -55 points. This Squad may have an M3 halftrack for +80 points or 1 Medium Truck for +50 points.

Support Assets

The following are all the Support Assets available to a US Army Armoured Infantry force.

HMG Company Support Team - 80 points

Supporting the platoon's advance, HMG teams suppressed enemies in cover, and annihilated any who tried to counterattack.

Type	Size	Move	Close Combat	Target	Save	Kill	Traits
Corporal	1	4"	D6	4+	6+	6+	—
Soldier	1	4"	D6	4+	6+	6+	—

Unit: Corporal (unit leader) with a M3 SMG, Soldier with a M2 .50 Cal HMG, Soldier with a M1 Garand rifle.

Options: Any member of the team may swap their M3 SMG or M1 Garand for an M1 Carbine for +2 points each. This Team may have an M3 halftrack for +80 points or a Jeep for +70 points.

81mm Mortar Company Support Team - 70 points

Laying down successive bombardments, mortar teams forced enemies to redeploy units in order to escape the worst effects of this light artillery.

Type	Size	Move	Close Combat	Target	Save	Kill	Traits
Corporal	1	4"	D6	4+	6+	6+	—
Soldier	1	4"	D6	4+	6+	6+	—

Unit: Corporal (unit leader) with a M3 SMG, Soldier with an 81mm Mortar (FaF), Soldier with a M1 Garand rifle.

Options: Any member of the team may swap their M3 SMG or M1 Garand for an M1 Carbine for +2 points each. This unit may have an M3 halftrack for +80 points or a Jeep for +70 points.

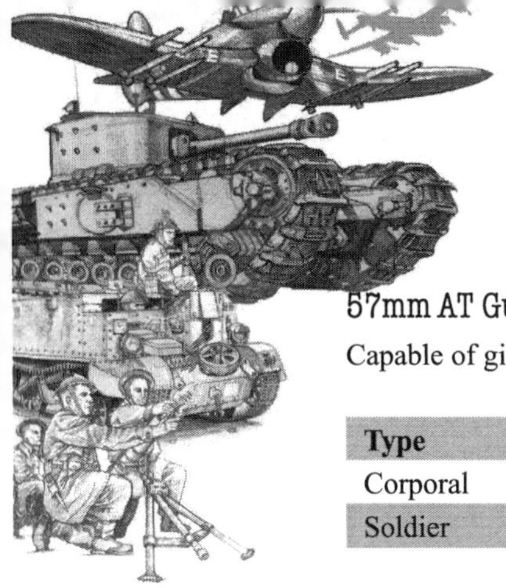

57mm AT Gun – 85 points

Capable of giving German armour pause for thought, anti-tank guns were vital in supporting their infantry.

Type	Size	Move	Close Combat	Target	Save	Kill	Traits
Corporal	1	4”	D6	4+	6+	6+	—
Soldier	1	4”	D6	4+	6+	6+	—

Unit: Corporal (unit leader) with a M3 SMG, Soldier a 57mm anti-tank gun (FaF), 2 Soldiers with M1 Garand rifles.

Crew Weapon: The 57mm anti-tank gun can only shoot if at least one crew member is in base contact with it.

Cumbersome: A Ready action must be taken before the 57mm anti-tank gun may be moved. The Move score of all models in the unit depends on the number of models still manning the weapon.

4 crew = 4” maximum move,
3 crew = 3” maximum move,
2 crew = 2” maximum move and
1 crew = No movement possible.

If the crew choose to abandon the 57mm Anti-Tank Gun, they may move normally.

Gun Shield: The Corporal and Soldiers count as being in Medium Cover when shot from the Forward facing.

Options: Any member of the team may swap their M3 SMG or M1 Garand for an M1 Carbine for +2 points each.

Combat Engineer Squad – 255 points

US combat Engineers served in every major US operation, and their job was often the most dangerous. It was their duty to do engineering work under enemy fire, take fortified buildings or clean mine fields.

Type	Size	Move	Close Combat	Target	Save	Kill	Traits
Sergeant	1	5”	D6+1	4+	6+	6+	—
Soldier	1	5”	D6+1	4+	6+	6+	—

Engineer Team: Sergeant (unit leader) with a M1 Carbine, 3 Soldiers with M1 Carbines.

BAR Team: Soldier with a M1918 BAR, Soldier with a M1 Garand rifle.

Bazooka Team: Soldier with a Bazooka and M1 Carbine, Soldier with a M1 Garand rifle.

Flamethrower Team: Soldier with a M2-2 Flamethrower, Soldier with a M1 Garand rifle.

Ambush: This unit may be concealed in an area of terrain outside the enemy’s deployment zone before the game begins. Note the location of the unit before any models are deployed on the table. You can reveal the ambush by placing the section on the table at any time after a unit (friendly or otherwise) has completed an action. However, the unit must be revealed by the end of turn three or be counted as destroyed.

Elite Training: The unit gains a +1 bonus to their Armour Saves when in Cover. They may also re-roll their Close Combat dice.

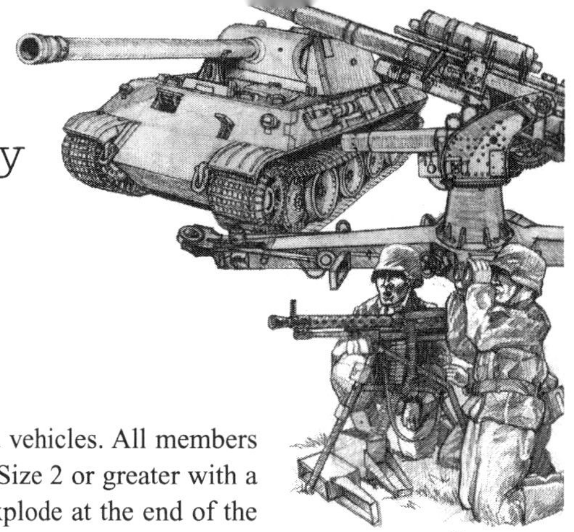

US Armoured Infantry

Fearless: The unit discards the first Suppression dice it suffers every turn.

Teams: One Soldier may act as unit leader to their team, splitting off to act as a separate unit.

Tank Assault: Infantry in this period used a variety of grenades or mines to assault armoured vehicles. All members of the unit are assumed to carry Anti-Tank Explosives. These may be placed on any model of Size 2 or greater with a Charge action if the target ends up within the Lethal Range of the charging unit. They will explode at the end of the charging unit's next turn.

Options: One Soldier may change his M2-2 Flamethrower for an M1919A6 .30 Cal HMG +40 points. The whole unit may be equipped with improved Anti-Tank Explosives for +30 points. This Squad may have an M3 halftrack for +80 points. If this option is taken, the Squad no longer benefits from the Ambush rule.

Late War US Armoury

Weapon	Range	Damage	**Traits**
57mm anti-tank gun	36"	D10+2	Lethal Zone/1", Ready, Slow
81mm Mortar	48"	D6	Artillery, Lethal Zone/3", Ready, Slow
Anti-Tank Explosives	Placed	D10	Lethal Zone/2", One-Shot, Piercing/2
Bazooka	18"	D10+2	Lethal Zone/ 1", Slow
Grenade Launcher	24"	D6	Lethal Zone/ 2", Piercing/1
Heavy Artillery	Special	2xD6+1	Lethal Zone/4", One-Shot
Improved Anti-Tank Explosives	Placed	2xD10+2	Lethal Zone/2", One-Shot, Piercing/3
Light Artillery	Special	D6	Lethal Zone/4", One-Shot
M1 Carbine	18"	2xD6	-
M1 Garand Rifle	24"	D6	Auto
M1918 BAR	24"	3xD6	Auto
M2 .50 Cal HMG	48"	2xD6+1	Auto, Piercing/1, Prone
M2-2 Flamethrower	10"	3xD6	Flame, Slow
M3 SMG	12"	3xD6	-
Scoped M1903A Springfield	36"	D6	Accurate

57mm anti-tank gun: Though outpaced towards the end of the war by new designs in German armour, few tank crews would willingly leave themselves open to sustained attack by a battery of these guns.

81mm Mortar: Though vulnerable to assault, mortar positions were quick to set up in battle, and provided a steady rain of fire on an increasingly dispirited enemy.

Anti-Tank Explosives: A combination of high-powered grenades and make-shift charges, these explosives give infantry a fighting chance against armoured vehicles, though their use is only advised in built-up areas.

Bazooka: With a shaped-charged warhead, the Bazooka finally gave infantry the means to destroy enemy tanks, and the design was to be copied by every major military force.

Grenade Launcher: Clipping onto the barrel of the M1 Garand rifle, this accessory allowed a soldier to launch grenades a good distance, providing an ad hoc mortar-like role.

US Armoured Infantry

Heavy Artillery: At the call of platoon leaders, heavy artillery could literally blast infantry out of cover, and was effective against all but the heaviest of armoured vehicles.

Improved Anti-Tank Explosives: Dedicated satchel charges were used by elite units in order to immobilise and destroy enemy armour, though such attacks were extremely dangerous.

Light Artillery: Using smaller and short-ranged guns, light artillery is primarily an anti-infantry weapon, forcing enemy troops into cover and disrupting their attacks as an assault is launched.

M1 Carbine: Based on the Garand, the M1 Carbine could be used in much the same way as a modern assault rifle.

M1 Garand Rifle: Replacing the Springfield as the US Army's main firearm, the Garand was an effective self-loading rifle.

M1918 BAR: A light machine gun, the BAR had selectable rates of fire, though was initially hampered by its small magazine.

M2 .50 Cal HMG: With integrated bipod, the Browning was an effective, though cumbersome, weapon.

M2-2 Flamethrower: Though to be made obsolete by advances in flame throwing tanks, the M2 was a vicious weapon in close quarters, and had significant advantages in built up areas.

M3 SMG: Nicknamed the grease gun, the M3 replaced the Thompson and was a good short-ranged weapon.

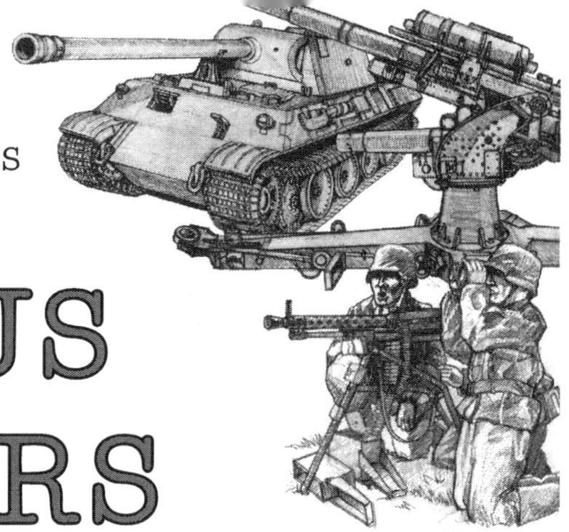

LATE WAR US PARATROOPERS

Airborne units were used in many European operations, including Operation Varsity, D-Day and Operation Market Garden. Once deployed paratrooper platoons would link up with the rest of their forces, before striking at enemy held positions, often miles behind the established front line. It could be a risky tactic, but the ability to send large numbers of infantry into territory that was deemed safe from attack by other means could tilt the balance for an entire campaign.

Platoon Structure

The standard squad size for most US Army platoons is twelve soldiers, consisting of two teams. The squad is led by a Sergeant, while the second team is led by a Corporal.

Three such sections make up a standard platoon, led by a Lieutenant, who is in turn assisted by a Sergeant. Three platoons make a rifle (infantry) company.

A platoon that has been on operational duty for some time may vary wildly from this, however. Heavy losses can result in fewer and smaller sections, while a platoon involved in heavy combat may become 'reinforced' with more soldiers, giving it more squads.

You can make any platoon with the US Paratrooper list, from one that has just been deployed, to a grizzled set of veterans that have taken heavy losses or been reinforced.

Each unit in a platoon is considered to be an Asset of some type. You may include the following Assets in your platoon, so long as you obey the minimum and maximum units listed for each. You may take one platoon for every 1,500 points in your force.

0-1 Command Asset
2-3 Squad Assets
0-3 Support Assets
0-1 Armour Asset per Squad Assets
0-3 Air Assets

For details on more Support, Armour and Air Assets see the US Vehicle and Air Assets List on page 134).

Special Rules

The following special rules apply to all US Paratrooper platoons.

Elite Training

All US Paratrooper infantry units gain a +1 bonus to their Armour Saves when in Cover. They may also re-roll their Close Combat dice.

Dedicated

Every US Paratrooper infantry unit rolls a D6 when Suppressed. If the result is equal to or less than the number of models currently in the unit, you can ignore all the effects of the Suppression.

Parachute Force Deployment

A US Paratroopers platoon may always deploy by parachute (see page 38). If you choose to do so, all units in the platoon capable of deploying this way must do so – any units that cannot deploy this way may not be used. The platoon may also take up to three Air Assets rather than two.

Command Assets

The following are all the Command Assets available to a US Paratrooper force.

Command Squad - 110 points

Forced to think on his feet as soon as he hits the ground, the Lieutenant carries a heavy responsibility for the safety of his men on the most dangerous of missions.

Type	Size	Move	Close Combat	Target	Save	Kill	Traits
Lieutenant	1	4”	D6	4+	6+	6+	-
Staff Sergeant	1	4”	D6	4+	6+	6+	-
Soldier	1	4”	D6	4+	6+	6+	-

Unit: Lieutenant with a M3 SMG, Staff Sergeant with a M3 SMG, Soldier with a M3 SMG, 4 Soldiers with M1 Garand rifles.

Supporting Artillery: This unit may purchase up to 3 uses of Supporting Artillery. Calling in Supporting Artillery requires a Ready action. This is used as an Artillery attack, but the initial Impact Point must be in the unit’s Line of Sight. There is no minimum range for this attack (yes, you can call it down right on your own head!). Light Artillery costs +50 points per use, while Heavy Artillery costs +100 points per use.

Options: The whole unit may be parachute deployed. Any member of the squad may swap their M3 SMG or M1 Garand for an M1 Carbine for +2 points each. You may add up to 2 LMG Teams for +60 points. The LMG Teams may be attached to the Command squad or may act independently.

LMG Team Command Attachment - 60 points

This small team can greatly enhance the capabilities of the entire platoon by laying down heavy firepower precisely where it is most needed.

Type	Size	Move	Close Combat	Target	Save	Kill	Traits
Soldier	1	4”	D6	4+	6+	6+	-

Unit: Soldier with a M1918 BAR, Soldier with a M1 Garand rifle.

Options: The whole unit may be parachute deployed. The M1 Garand may be swapped for an M1 Carbine for +2 points.

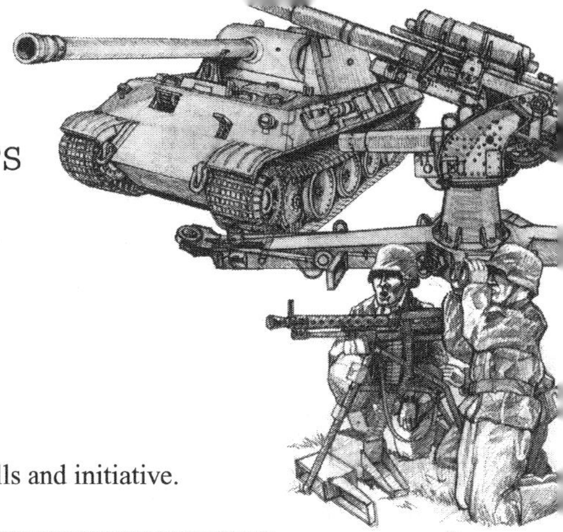

Squad Assets

The following are all the Squad Assets available to a US Paratrooper force.

Paratrooper Squad - 220 points

Dropping into enemy territory, paratroopers are expected to act with a reliance on their own skills and initiative.

Type	Size	Move	Close Combat	Target	Save	Kill	Traits
Sergeant	1	4”	D6	4+	6+	6+	-
Corporal	1	4”	D6	4+	6+	6+	-
Soldier	1	4”	D6	4+	6+	6+	-

Rifle Team: Sergeant with a M3 SMG, 7 Soldiers with M1 Garand rifles.

MG Team: Corporal with M3 SMG, Soldier with M1918 BAR, 2 Soldiers with M1 Garand rifle.

Teams: The Corporal may act as unit leader to the MG Team, splitting off to act as a separate unit.

Tank Assault: Infantry in this period used a variety of grenades or mines to assault armoured vehicles. All members of the unit are assumed to carry Anti-Tank Explosives. These may be placed on any model of Size 2 or greater with a Charge action if the target ends up within the Lethal Range of the charging unit. They will explode at the end of the charging unit’s next turn.

Options: The whole unit may be parachute deployed. One Soldier in the Rifle Team may be equipped with a Bazooka in addition to his rifle for +20 points. One Soldier in the Rifle Team may swap his M1 Rifle for an M1918 BAR for +30 points. One Soldier per team may swap his M1 Garand for an M1 Carbine for +2 points each. One Soldier may swap his M1 Rifle for a Scoped M1903A Springfield for +10 points. Drop the MG Team for -80 points.

Support Assets

The following are all the Support Assets available to a US Paratrooper force.

Paratrooper 81mm Mortar Support Team - 130 points

Laying down successive bombardments, mortar teams forced enemies to redeploy units in order to escape the worst effects of this light artillery.

Type	Size	Move	Close Combat	Target	Save	Kill	Traits
Corporal	1	4”	D6	4+	6+	6+	-
Soldier	1	4”	D6	4+	6+	6+	-

Unit: Corporal with a M3 SMG, Soldier with an 81mm Mortar (FaF), 4 Soldiers with M1 Garand rifles.

US Paratroopers

Options: The whole unit may be parachute deployed. Any member of the team may swap their M3 SMG or M1 Garand for an M1 Carbine for +2 points each.

Paratrooper HMG Support Team – 90 points

Supporting the platoon's advance, HMG teams suppressed enemies in cover, and annihilated any who tried to counterattack.

Type	Size	Move	Close Combat	Target	Save	Kill	Traits
Corporal	1	4"	D6	4+	6+	6+	-
Soldier	1	4"	D6	4+	6+	6+	-

Unit: Corporal with a M3 SMG, Soldier with a M2 .50 Cal HMG, Soldier with a M1 Garand rifle.

Options: The whole unit may be parachute deployed. Any member of the team may swap their M3 SMG or M1 Garand for an M1 Carbine for +2 points each.

Paratrooper Pathfinder Team – 70 points

The Pathfinders calls in accurate artillery fire wherever it is needed, and are usually the first airborne troops to hit the ground.

Type	Size	Move	Close Combat	Target	Save	Kill	Traits
Sergeant	1	4"	D6	4+	6+	6+	-
Soldier	1	4"	D6	4+	6+	6+	-

Unit: Sergeant with a M3 SMG, 2 Soldiers with M1 Carbines, Soldier with a M1 Garand rifle.

Supporting Artillery: This unit may purchase up to 1 uses of Supporting Artillery. Calling in Supporting Artillery requires a Ready action. This is used as an Artillery attack, but the initial Impact Point must be in the unit's Line of Sight. There is no minimum range for this attack (yes, you can call it down right on your own head!). Light Artillery costs +50 points per use, while Heavy Artillery costs +100 points per use.

Options: The whole unit may be parachute deployed. Any member of the team may swap their M3 SMG or M1 Garand for an M1 Carbine for +2 points each.

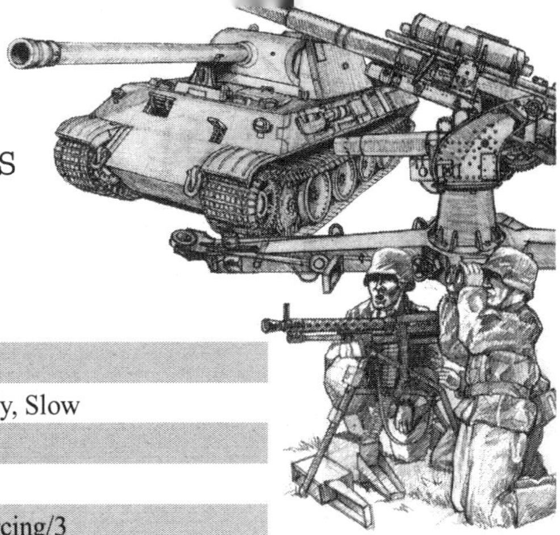

Late War US Armoury

Weapon	Range	Damage	**Traits**
81mm Mortar	48"	D6	Artillery, Lethal Zone/3", Ready, Slow
Bazooka	18"	D10+2	Lethal Zone/ 1", Slow
Heavy Artillery	Special	2xD6+1	Lethal Zone/4", One-Shot
Improved Anti-Tank Explosives	Placed	2xD10+2	Lethal Zone/2", One-Shot, Piercing/3
Light Artillery	Special	D6	Lethal Zone/4", One-Shot
M1 Carbine	18"	2xD6	-
M1 Garand Rifle	24"	D6	Auto
M1918 BAR	24"	3xD6	Auto
M2 .50 Cal HMG	48"	2xD6+1	Auto, Piercing/1, Prone
M3 SMG	12"	3xD6	-
Scoped M1903A Springfield	36"	D6	Accurate

81mm Mortar: Though vulnerable to assault, mortar positions were quick to set up in battle, and provided a steady rain of fire on an increasingly dispirited enemy.

Bazooka: With a shaped-charged warhead, the Bazooka finally gave infantry the means to destroy enemy tanks, and the design was to be copied by every major military force.

Heavy Artillery: At the call of platoon leaders, heavy artillery could literally blast infantry out of cover, and was effective against all but the heaviest of armoured vehicles.

Improved Anti-Tank Explosives: Dedicated satchel charges were used by elite units in order to immobilise and destroy enemy armour, though such attacks were extremely dangerous.

Light Artillery: Using smaller and short-ranged guns, light artillery is primarily an anti-infantry weapon, forcing enemy troops into cover and disrupting their attacks as an assault is launched.

M1 Carbine: Based on the Garand, the M1 Carbine could be used in much the same way as a modern assault rifle.

M1 Garand Rifle: Replacing the Springfield as the US Army's main firearm, the Garand was an effective self-loading rifle.

M1918 BAR: A light machine gun, the BAR had selectable rates of fire, though was initially hampered by its small magazine.

M2 .50 Cal HMG: With integrated bipod, the Browning was an effective, though cumbersome, weapon.

M3 SMG: Nicknamed the grease gun, the M3 replaced the Thompson and was a good short-ranged weapon.

Scoped M1903A Springfield: Though replaced by the Garand, the Springfield found plenty of life in the late war period as a scoped sniper rifle.

LATE WAR US VEHICLE AND AIR ASSETS

The following Assets may be used by any Late War US Army force.

Support Assets

The following are all the Support Assets available to a US Army force.

Jeep – 70 points

Credited with making victory possible, the Jeep was a very basic design, but all the more rugged and capable for that.

Type	Size	Move	Close Combat	Target	Save	Kill	Traits
Jeep	2	8”	D10	6+	6+	7+	Hits/2, Lumbering, Wheels/15”

Unit: Jeep with a Browning .50 cal AAMG (FaF).

Transport: The Jeep can carry up to 5 Size points of models (each maximum Size 1).

M16 AA MGMC – 160 Points

A variant of the M3 Halftrack, this vehicle sported a quad array of M2 Browning MGs on an anti-aircraft mount, making it a fearsome prospect for any Axis pilot.

Type	Size	Move	Close Combat	Facing	Target	Save	Kill	Traits
M16 AA MGMC	3	7	2xD10	Front	7+	5+	8+	Hits/2, Lumbering, Tracks/14
				Sides	6+	6+	7+	
				Rear	6+	6+	7+	

Unit: M16 with a Browning .50 cal AAMG quad mount (FaF).

M20 Scout Car - 135 points

Based on the M8, this variant had its turret removed and replaced with an anti-aircraft mount for an M2 machine gun.

Type	Size	Move	Close Combat	Facing	Target	Save	Kill	Traits
M20 Scout Car	3	10"	2xD10	Front	7+	3+	9+	Hits/2, Lumbering, Wheels/19"
				Sides	7+	5+	9+	
				Rear	6+	5+	8+	

Unit: M20 with a Browning .50 cal AAMG.

Transport: The M20 can carry up to 4 Size points of models (each maximum Size 1).

M3 Halftrack - 80 points

With dozens of variants made, the M3 was a truly versatile vehicle, and many were sent to allied countries as part of the Lend-Lease agreement.

Type	Size	Move	Close Combat	Facing	Target	Save	Kill	Traits
M3 Halftrack	3	7"	2xD10	Front	7+	5+	8+	Hits/2, Lumbering, Tracks/14"
				Sides	6+	6+	7+	
				Rear	6+	6+	7+	

Unit: M3 Halftrack with a Pintle MG (FaF).

Transport: The M3 Halftrack can carry up to 12 Size points of models (each maximum Size 1).

Options: A Pintle MG (FaA) may be added for +35 points.

M8 Armoured Car - 175 points

Fast and lightly armoured, the M8 was often used in the reconnaissance role, though it also served as a light tank destroyer.

Type	Size	Move	Close Combat	Facing	Target	Save	Kill	Traits
M8	2	10"	2xD10	Front	7+	3+	9+	Hits/2, Lumbering, Tracks/19"
				Sides	7+	5+	9+	
				Rear	6+	5+	8+	

Unit: M8 with a M6 37mm gun and a Browning .50 cal AAMG.

US Vehicles

Medium Truck - 50 points

A variety of trucks were used by the US throughout the war, and they formed a vital logistical link between armies in the field and their supplies.

Type	Size	Move	Close Combat	Target	Save	Kill	Traits
Medium Truck	4	7"	D10	6+	6+	7+	Hits/3, Lumbering, Wheels/14"

Unit: Medium Truck.

Transport: The Medium Truck can carry up to 10 Size points of models (each maximum Size 1).

Armour Assets

The following are all the Armour Assets available to a US Army force.

M10 Wolverine Tank Destroyer - 195 points

Intending to use speed and agility to combat German tanks, the M10 Wolverine was an interim step between the inadequate tanks used earlier in the war and the next generation.

Type	Size	Move	Close Combat	Facing	Target	Save	Kill	Traits
M10	4	4"	4xD10	Front	7+	4+	11+	Hits/2, Lumbering, Tracks/8"
				Sides	7+	5+	8+	
				Rear	7+	5+	7+	

Unit: M10 with an M7 anti-tank gun and a Browning .50 cal AAMG.

Browning .50 cal AAMG: This weapon may not be fired with the same Shoot action that the M7 anti-tank gun is fired.

M18 Hellcat Tank Destroyer - 165 points

Extremely fast (hence the name Hellcat), this tank destroyer had relatively weak armour, and so had to be used by tactically-minded crew when engaging enemy tanks.

Type	Size	Move	Close Combat	Facing	Target	Save	Kill	Traits
M18	4	8"	3xD10	Front	7+	5+	10+	Hits/2, Lumbering, Tracks/16"
				Sides	7+	6+	8+	
				Rear	7+	6+	7+	

Unit: M18 with an M7 anti-tank gun and a Browning .50 cal AAMG

Browning .50 cal AAMG: This weapon may not be fired with the same Shoot action that the M7 anti-tank gun is fired.

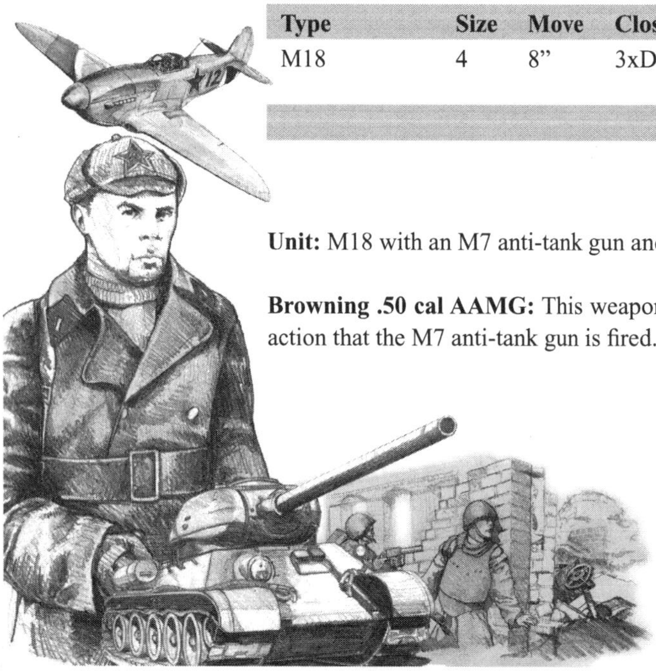

M24 Chaffee Light Tank – 185 points

A light tank, the M24 was designed to be faster and cheaper to produce than the Sherman. It could not stand toe-to-toe with heavy German armour, but was extremely capable when supporting infantry.

Type	Size	Move	Close Combat	Facing	Target	Save	Kill	Traits
M24 Chaffee	3	5"	2xD10	Front	8+	5+	10+	Hits/2, Lumbering, Multifire/2, Tracks/11"
				Sides	7+	6+	9+	
				Rear	7+	6+	8+	

Unit: M24 Chaffee with a M6 75mm gun, a Browning .50 cal AAMG, and an MG (FaF) and an MG.

Browning .50 cal AAMG: This weapon may not be fired with the same Shoot action that the M6 75mm gun is fired.

Co-Axial MG: To improve target acquisition, MGs on vehicles were often mounted co-axial to the main gun. One MG has a co-axial mount. The co-axial MG must always use the same Fire Zone as the main gun. As soon as a Fire Zone is placed within range of the co-axial MG, you may add +1 to all Damage Dice of the main gun.

M4A1(76) W Sherman Medium Tank – 245 points

Intended to counter the threat of heavy German armour, this variant of the Sherman featured the M7 anti-tank gun, a weapon capable of blasting apart the most heavily protected enemy tanks.

Type	Size	Move	Close Combat	Facing	Target	Save	Kill	Traits
M4A1 (76) W Sherman	4	3"	4xD10	Front	8+	4+	12+	Hits/2, Lumbering, Multifire/2, Tracks/7"
				Sides	7+	5+	11+	
				Rear	7+	5+	10+	

Unit: M4A3 Sherman with an M7 anti-tank gun, a Browning .50 cal AAMG, an MG (FaF) and an MG.

Browning .50 cal AAMG: This weapon may not be fired with the same Shoot action that the M7 anti-tank gun is fired.

Co-Axial MG: To improve target acquisition, MGs on vehicles were often mounted co-axial to the main gun. One MG has a co-axial mount. The co-axial MG must always use the same Fire Zone as the main gun. As soon as a Fire Zone is placed within range of the co-axial MG, you may add +1 to all Damage Dice of the main gun.

US Vehicles

M4A3 Sherman Medium Tank - 225 points

Used in great numbers by both the US and its allies, the Sherman was a solid and reliable tank but its main gun had trouble penetrating the armour of late war German tanks.

Type	Size	Move	Close Combat	Facing	Target	Save	Kill	Traits
M4A3 Sherman	4	3"	4xD10	Front	8+	4+	12+	Hits/2, Lumbering, Multifire/2, Tracks/7"
				Sides	7+	5+	11+	
				Rear	7+	5+	10+	

Unit: M4A3 Sherman with a M3 75mm gun, a Browning .50 cal AAMG, an MG (FaF) and an MG.

Browning .50 cal AAMG: This weapon may not be fired with the same Shoot action that the M7 anti-tank gun is fired.

Co-Axial MG: To improve target acquisition, MGs on vehicles were often mounted co-axial to the main gun. One MG has a co-axial mount. The co-axial MG must always use the same Fire Zone as the main gun. As soon as a Fire Zone is placed within range of the co-axial MG, you may add +1 to all Damage Dice of the main gun.

M4A3E2 (76) Jumbo Sherman Medium Tank - 480 points

Designed to attack fortifications, the Jumbo Sherman sported heavier armour and a redesigned turret.

Type	Size	Move	Close Combat	Facing	Target	Save	Kill	Traits
M4A3E2 (76) Sherman	4	3"	4xD10	Front	9+	3+	13+	Hits/3, Lumbering, Multifire/2, Tracks/7"
				Sides	9+	4+	13+	
				Rear	8+	4+	12+	

Unit: M4A3E2 (76) Sherman with an M7 anti-tank gun, a Browning .50 cal AAMG, an MG (FaF) and an MG.

Browning .50 cal MG: This weapon may not be fired with the same Shoot action that the M7 anti-tank gun is fired.

Co-Axial MG: To improve target acquisition, MGs on vehicles were often mounted co-axial to the main gun. One MG has a co-axial mount. The co-axial MG must always use the same Fire Zone as the main gun. As soon as a Fire Zone is placed within range of the co-axial MG, you may add +1 to all Damage Dice of the main gun.

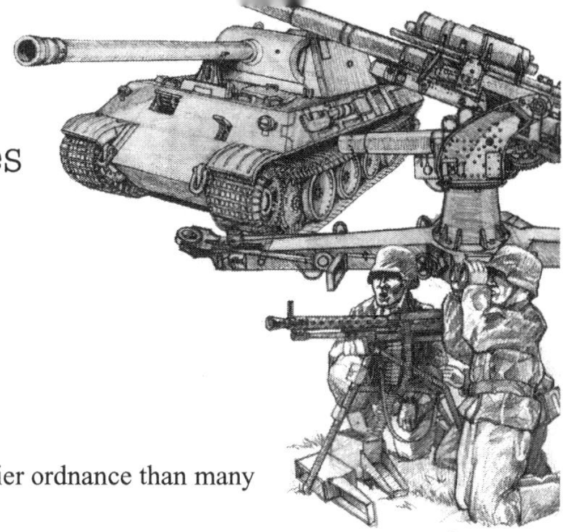

Air Assets

The following are all the Air Assets available to a US Army force.

P-47D Thunderbolt - 270 points

A good fighter in the air, the Thunderbolt excelled in the ground attack role. Able to carry heavier ordnance than many other fighter-bombers, it could devastate an enemy attack before momentum had been built up.

Type	Size	Move	Close Combat	Target	Save	Kill	Traits
P-47D Thunderbolt	5	Loiter Cruise	4xD10	6+	5+/5+	8+	Hits/3, Multifire/6

Unit: P-47D Thunderbolt with six Browning M2 MGs (FaF) and two 1000 lb. Bombs.

1,000 lb. Bombs: The P-47D Thunderbolt may drop up to 2 Bombs with every Shoot (bomb) action.

US Vehicle Armoury

Weapon	Range	Damage	Traits
1,000 lb. Bomb	Dropped	D10+2	Lethal Zone/6", Multihit, One Shot
Browning .50 cal AAMG	30"	3xD6	AA, Auto, Piercing/1
Browning .50 cal AAMG quad mount	30"	6xD6	AA, Acurate, Auto, Piercing/1
Browning M2 MG	20"	D6+1	Auto, Piercing/1
M3 75mm gun	48"	D10+2	Lethal Zone/1", Slow
M6 37mm gun	48"	D6+3	Lethal Zone/2", Slow
M6 75mm gun	48"	D6+3	Lethal Zone/2", Piercing/1, Slow
M7 anti-tank gun	60"	D10+3	Lethal Zone/1", Piercing/1, Slow
MG	24"	3xD6	
Pintle MG	24"	3xD6	AA

1,000 lb. Bomb: A heavy bomb used by fighter-bombers and attack craft, accurate delivery could destroy armoured vehicles and tear apart bunkers,

Browning .50 cal AAMG: One of the most widely used machine guns in the world, the Browning can be found on vehicles and aircraft throughout the 20th Century, and was very common as the main armament on Allied aircraft.

Browning .50 cal AAMG quad mount: The M16 halftrack was able to bear an incredible amount of firepower upon enemy aircraft by using four Browning machine guns in a single mount.

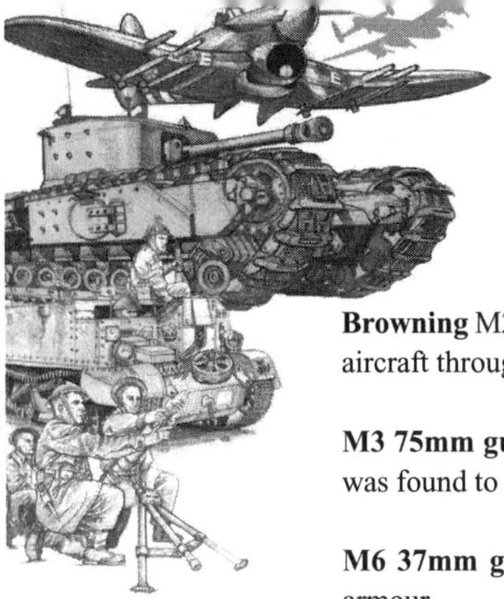

US Vehicles

Browning M2 MG**:** One of the most widely used machine guns in the world, the Browning can be found on vehicles and aircraft throughout the 20th Century, and was very common as the main armament on Allied aircraft.

M3 75mm gun: The standard gun of the Sherman, this main gun stood in good stead against early German tanks, but was found to be lacking in the later periods of the war.

M6 37mm gun: Useful for blasting dug-in infantry out of heavy cover, the M6 was not suited to engaging heavy armour.

M6 75mm gun: Mounted on light tanks, this gun could damage heavy armour, but had great trouble making account of itself unless specific weak points could be targeted.

M7 anti-tank gun: Introduced to US forces in order to counter heavy German armour, the M7 proved to be successful in maintaining the balance in the race for heavy armour between Axis and Allies.

MG: Covering a wide variety of weapons, machine guns mounted on vehicles are used to suppress enemy infantry and provide supporting fire.

Pintle MG: When seated on a pintle mount, a machine guns gains a greater field of fire and elevation, allowing it to be used to engage low-flying aircraft.

ENGAGEMENTS

This chapter deals with creating scenarios for Battlefield Evolution. An engagement is the default type of battle in this game and consists of two balanced forces on the battlefield, each with a mission. This mission is what gives a player his goals for an engagement and will therefore be of key importance to his strategy. Both players will need to know the following information in order to generate an engagement:

- The agreed Force Value, which represents how large a force you are using. Both you and your opponent will have the same amount of points to spend on your army.
- What particular tactics you wish to use (Attacking, Defending or Probing).

Force Value

Force Value is used to ensure that both sides have an equivalent strength in the battle. All units in the Battlefield Evolution are allotted a Value to assist in constructing forces. It is customary for players to agree to a pre-arranged total points value for each force (referred to as the Force Value). The totalled Value characteristics of all units in your army, including any optional benefits, cannot exceed this Force Value. Common Force Values are:

- 500 to 1,000 points for small, skirmish games. These small Force Values are often used early in Battlefield Evolution: World at War to come to grips with the rules and systems.
- 1,000-3,000 points is the range in which most games of Battlefield Evolution: World at War are played – this produces a decent sized game with a variety of different units in play. Most players aim for Force Values in this range when constructing their forces.
- More than 3,000 points makes for a highly involved and varied engagement, usually involving multiple platoons and huge armies.

Tactics

Your tactics are part of your battle plan and you decide on them before you deploy, giving you a chance to look at the battlefield and know which army you are fighting before committing to a course of action. By matching your tactics against those of your enemy, the type of mission you will be fighting will be determined.

Your choices of tactics are;

Attack: Throw your force into a thrust on a narrow frontage, sacrificing mobility to put pressure on a critical point.

Defend: Aim to avoid close encounters and put emphasis on a tough deployment.

Probe: Adopt a more mobile and flexible approach, able to fight battles on the spur of the moment.

Once you have chosen your tactics, you should consult the Mission Table to find out what your specific orders are.

Both you and your opponent should be honest about what missions you are using, as knowing what the enemy is up to will prove to be very important. It is likely that you will be able to win a battle as much by stopping the enemy achieving his objectives as by achieving your own.

You will earn bonus Victory Points (see page 40) for achieving your mission objectives.

Mission Table

Opponent's Tactics \ Your Tactics	*Attack*	*Defend*	*Probe*
Probe	Overrun	Delay	Hold
Defend	Overrun	Hold	Overrun
Attack	Annihilate	Last Stand	Delay

Missions

The following descriptions will list the mission you must attempt and what bonus Victory Points can be earned by completing its objectives.

Annihilate

Your orders are for you to rack up the body count. You have been commanded to eradicate everything you can. Anything you cannot kill, at least soften up so someone else can come along behind you and finish the job.

Bonus Victory Points: Earn a +25% bonus on all Victory Points earned for completely wiping out an enemy unit.
Penalty: Lose all bonus Victory Points if no unit in your force is above half strength.

Delay

It is short straw time for you and your command. You have been issued orders to delay the enemy at any cost. Your casualties are irrelevant; engage their attention and inflict casualties wherever and whenever you can but do not let them overrun your own positions in the process.

Bonus Victory Points: If the Game ends and you still have at least one unit that has not been reduced to half strength, +25% to your Victory Points total.
Penalty: Deduct the full Value of every enemy unit that is within your deployment area at the end of the game, regardless of its remaining strength in models.

Hold

Do not quit, it is as simple as that. Engage the enemy if you have to but do not give up an inch of ground doing it. Crush them if you can and do not get too badly hurt in the process. How hard can it be?

Bonus: Full value of every unit in your starting force that survives above half strength.
Penalty: Deduct half the Value of every enemy unit that is outside their deployment area at the end of the game, regardless of its remaining strength in models.

Last Stand

Just stay alive as long as you can. Reinforcements may or may not be coming but it does not matter either way. The enemy is closing in and one way or another you are not going to make victory cheap for them. Fight, kill and stay alive.

Special rules: Divide your force into two groups with at least a third of your total number of units in one group. Randomly determine which part will be deployed at the beginning of the game. The other force becomes available as Reserves at the beginning of your third turn.

Bonus: You will receive the full Value of every friendly unit that survives the game above half strength.
Penalty: Lose full Value of every friendly unit completely wiped out.

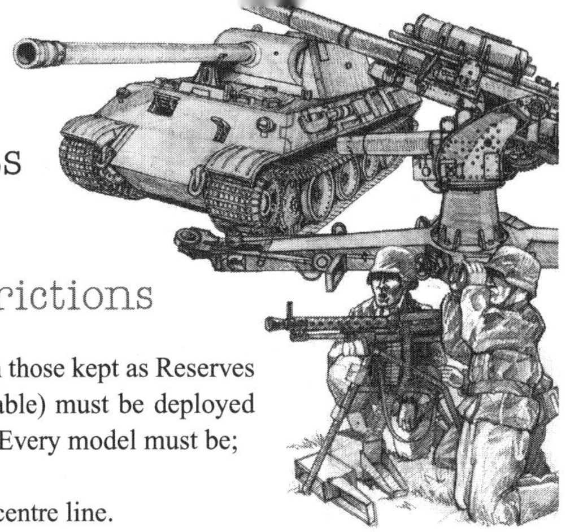

Overrun

The enemy force are just screening a tactical advantage behind their lines. They need to be knocked aside and overrun to gain ground. Victory is just beyond that next hill; unfortunately there are a lot of enemy troops between you and it.

Bonus: Full value of every unit you have in the enemy deployment area at the end of the game regardless of its remaining strength in models.
Penalty: Deduct the full value of every one of your units that still has models within your deployment area at the end of the game.

Deployment

Once you have determined which missions you and your opponent will be using, it is time to see what kind of deployment zones you will be using. These can radically affect the type of battle you fight, and the player who can select the ground on which he will fight will have an advantage.

The player who chooses what type of deployment you will use and which side of the table each army will set up is based on the tactics selected. Consult the Battle Advantage Chart below. The player with the highest level of tactics will choose which type of deployment will be used and on which side of the table he will deploy his army.

Battle Advantage

Defend
Probe
Attack

If both players have chosen the same tactics, each should roll a D10. The highest rolling player will have the Battle Advantage, and can choose the deployment type and where he will deploy his army.

Deployment Zones

The three types of deployment you can choose are Battle Line, Deep Defence and Encounter. Each of these is defined by where the centreline of the table runs, as shown on the diagrams below. This ensures any size of table can be used in Battlefield Evolution: World at War.

Deployment Restrictions

All models in an army (other than those kept as Reserves or otherwise not placed on the table) must be deployed within its own deployment zone. Every model must be;

- More than 12" from any centre line.
- Within 6" of its unit leader.

Who Goes First?

The player who did not get to choose the deployment areas must now choose whether he should set up his army first and take the first turn, or whether he should force his opponent to do so.

Tactical Advantages

Attack and Probe tactics have their own advantage beyond influencing deployment zones and who goes first.

Attack: This tactic allows an army to keep up to half of its unit in Reserve (see page 39).
Probe: This gives the option keeping up to half an army's units off the table until their first turn, instead of deploying them at the start of the battle. Units may move on from any table edges of the player's deployment area but they may not move on within 6" of any centre line or 12" of any enemy model.

Game Length

Using the Engagements system, games will go on for six turns.

Winning the Game

Players total their Victory Points (see page 40) at the end of the game, using the criteria given in their mission description. The player with the most Victory Points wins.

For the purposes of bonus points, units that are airborne do not count as being within a deployment area.

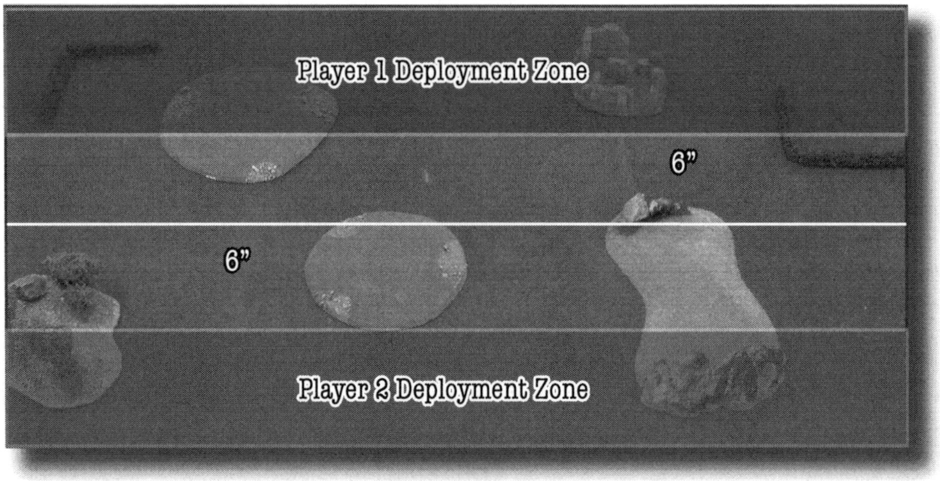

Battleline

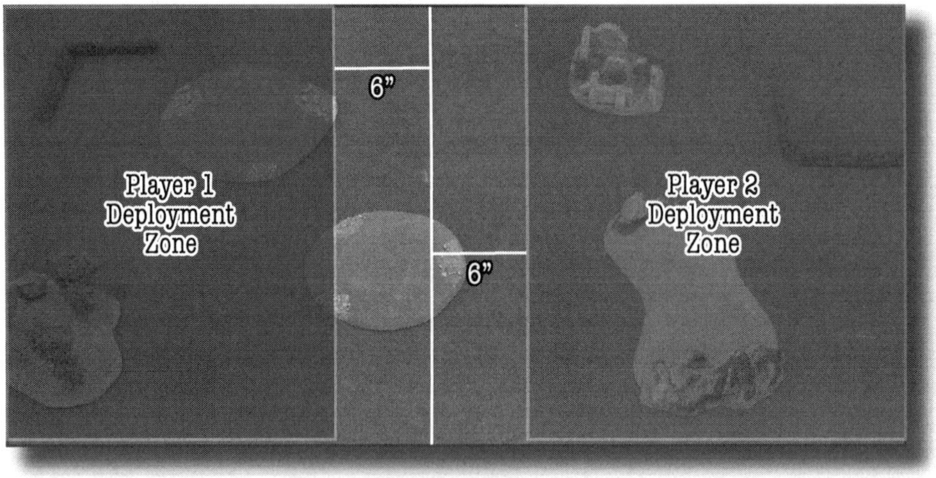

Deep Defence

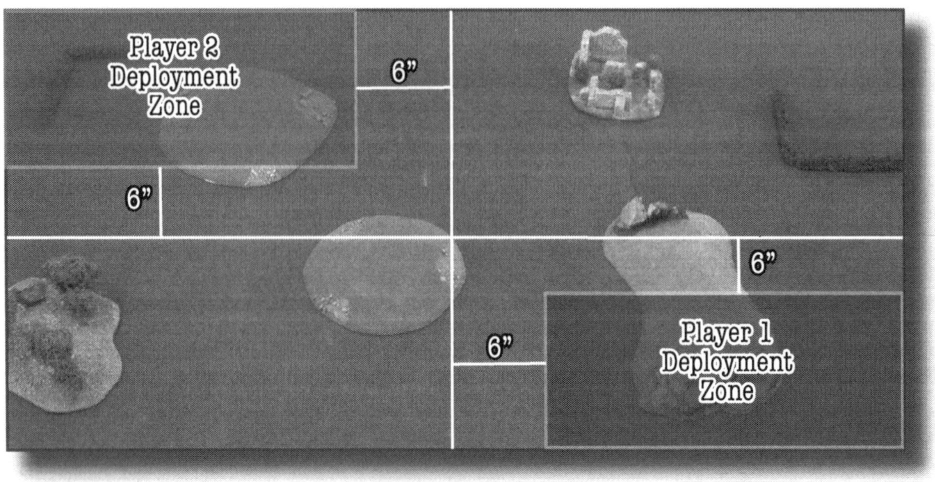

Encounter

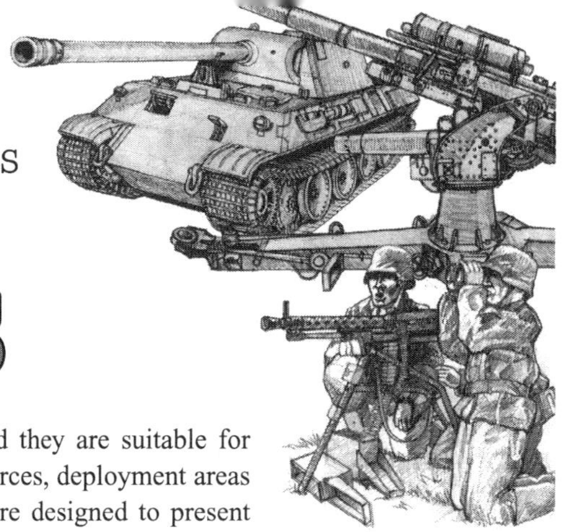

SCENARIOS

The Engagements chapter allows you to generate generic battles rather than scenarios, and they are suitable for tournaments and competitive games. A scenario is a pre-selected engagement that details the forces, deployment areas and/or victory conditions for both sides without using the Engagements system. Scenarios are designed to present specific tactical challenges, and are not always strictly fair!

CLASH OF ARMS

Two evenly matched forces meet on the frontline. Both sides have been building up to hostilities and each is eager to score the first victory in what is likely to be a long and bloody war.

Briefing

A misjudgement in the tactical prowess of your enemy has left the door wide open, and a strike force is on its way to deliver the hammer blow to your forward base of operations. Rapidly ordering units to the frontline, you must hold back this attack long enough for stronger lines of defence to be drawn up further down the road.

- Both players roll a dice. The lowest scoring player deploys his army first and takes the first turn.
- Reserves may not be used.

Scenario Rules

The scenario will last until the Victory Conditions are met.

Victory Conditions

The game ends when one army reduces the other to less than a quarter of its starting models in order to declare victory.

Player 1 Deployment Zone

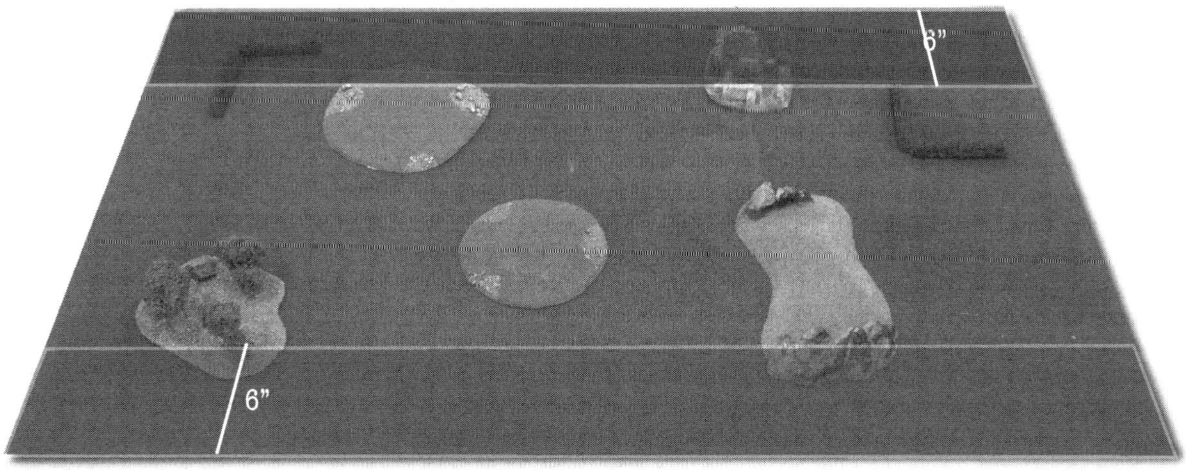

Player 2 Deployment Zone

FORMATION AMBUSH

An advancing line of troops has unwittingly ventured too far into enemy territory without deploying adequate scouts. Waiting for just such an error in tactical judgement, the opposing army lies in wait, ready to unleash a vicious attack to destroy the formation before it can adequately attack.

Attacker's Briefing

You have timed the ambush to perfection, and grin as the enemy advances, completely unaware of the massed firepower arrayed against him. Though you are attacking a larger force, you have claimed the element of surprise utterly, and can rely on massed confusion within the enemy formation to aid you in the battle.

- You may not include any Air Units.
- You may rearrange any terrain in your deployment zone as you see fit.
- The total points value of your force may be no greater than half that of the attacker's.
- You deploy your entire army second and take the first turn.

Defender's Briefing

A simple mission advancing into supposedly secure territory has turned into a nightmare. As soon as you heard the first chatter of fire, you knew you were in trouble – but can the rest of your force react as quickly?

- You may not include any Air Units.
- You deploy your entire army first and take the second turn.

Scenario Rules

The scenario will last for six turns. None of the defender's units may take any actions in their first turn. In the defender's second turn, roll a dice for each unit. On an even number, the unit may act normally. On an odd number, it may only take Move actions.

Victory Conditions

The army that earns the most Victory Points will win the battle.

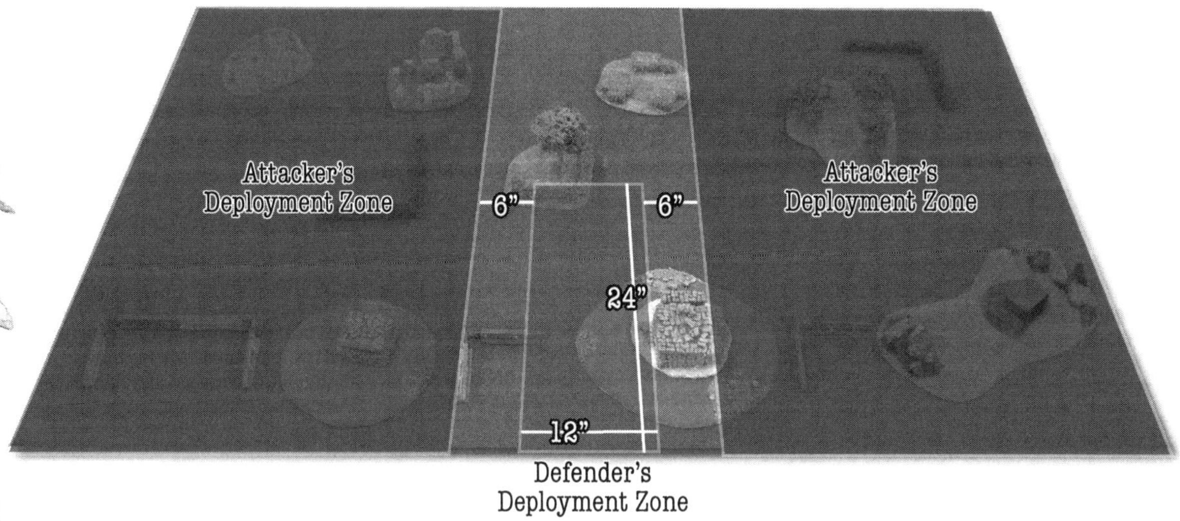

HAMMER AND ANVIL

A line of tightly integrated defences poses a serious tactical problem to both armies in this clash of arms. One side must contend with a deeply-rooted enemy who has no wish to relinquish territory, while the other is forced to adopt a static posture against a manoeuvring enemy, trusting to the security of armoured walls.

Attacker's Briefing

The enemy's forward post lies before you. One well timed, devastating attack here will decapitate his command structure, leaving the entire region in disarray and unable to fight effectively. Win this battle, and the entire war may be yours!

- You deploy your entire army first and take the second turn.

Defender's Briefing

You have been tasked with defending the command post with your life, as it is the last anchor point for the whole front line. Hold your position no matter what the cost, and make the enemy pay for every inch he tries to cover while advancing towards you.

- You must include one structure of any type in your deployment zone for every 500 points spent on your force.
- You deploy your entire army second and take the first turn.

Scenario Rules

The scenario will last for eight turns.

Victory Conditions

Victory Points are used in this scenario. The attacker gains +100 bonus Victory Points for every structure that he destroys. The defender gains bonus Victory Points equal to the Value of every surviving unit he has after turn eight that is within a structure, no matter what losses it has sustained.

Attacker's Deployment Zone

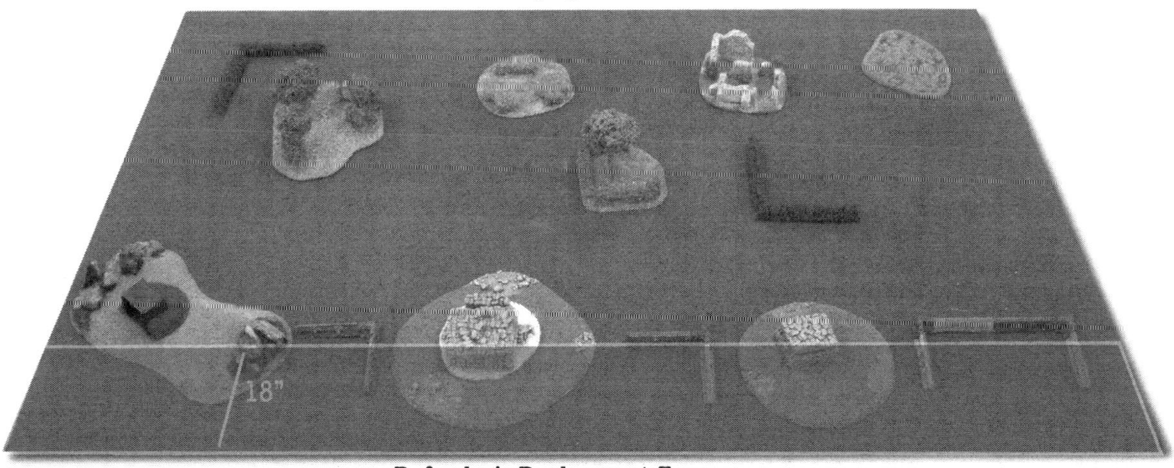

Defender's Deployment Zone

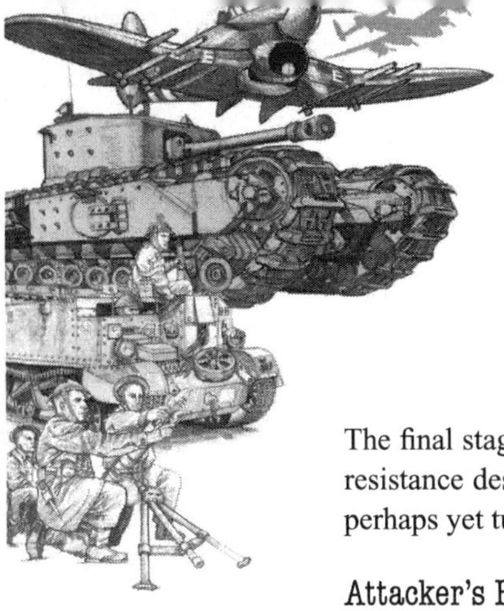

FINAL DEFENCE

The final stages of war are evident, and the defender has been almost completely defeated. Here and there, pockets of resistance desperately try to stem the flow of enemy forces, hoping against hope for a miracle that will save them and perhaps yet turn the tide of war.

Attacker's Briefing

The enemy is almost finished! Just clear this last position and they will be in full flight with nowhere to run. Show no mercy! Final victory is at hand!

- You deploy your entire army first and take the second turn.

Defender's Briefing

The war has gone hard and all hope is gone. All you can do is fight against an enemy that will accept no quarter and pray that you can buy enough time, perhaps for reinforcements to arrive, perhaps to just to allow your allies to regroup without interference for their own last stand. Maybe all you can do is buy a vital hour to let a few civilians evacuate a nearby city before the enemy rolls through their homes.

- The total points value of your force may be no greater than half that of the attacker's.
- You may arrange the terrain in your deployment zone however you wish, and may add more terrain as you see fit.
- You may include one structure of any type in your deployment zone for every 750 points spent on your force.
- You deploy your entire army second and have the first turn.

Scenario Rules

The scenario will last for six turns.

Victory Conditions

Determining the victor is simple in this scenario. If the defender has any units left at the end of the sixth turn, no matter how badly damaged, he may claim a (moral) victory. If the defender is wiped out before then, the attacker wins.

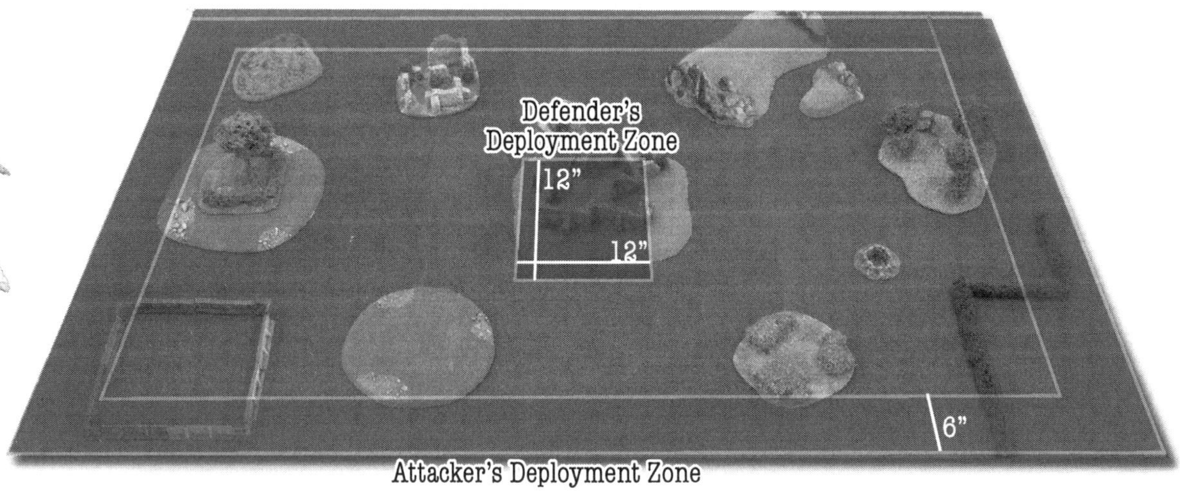

SEIZE AND SECURE

A tactically important landmark is being fought over by two armies. Whoever gains the upper hand for just a few hours will gain an advantage over the entire region that will be difficult to challenge. Just a little extra effort now will pay dividends later on.

Briefing

Study of the surrounding area yielded a variety of tactical options open to you, if only the region could be secured. Dispatching a force to take the objective, it was with some dismay that you saw the enemy approaching in force as well – clearly someone within the enemy command structure had the same thought and now you must fight over this scrap of land.

- Place a terrain piece or structure in the middle of the table (a hill or a building work well for this). This is the tactical objective you are fighting for.
- Roll a dice each – the lowest score will deploy first and take the first turn.

Scenario Rules

The scenario will last for six turns.

Victory Conditions

Victory Points are used in this scenario. Both sides gain bonus Victory Points equal to the full Value of each of their units that are within 12” of the objective, regardless of how many casualties they have taken.

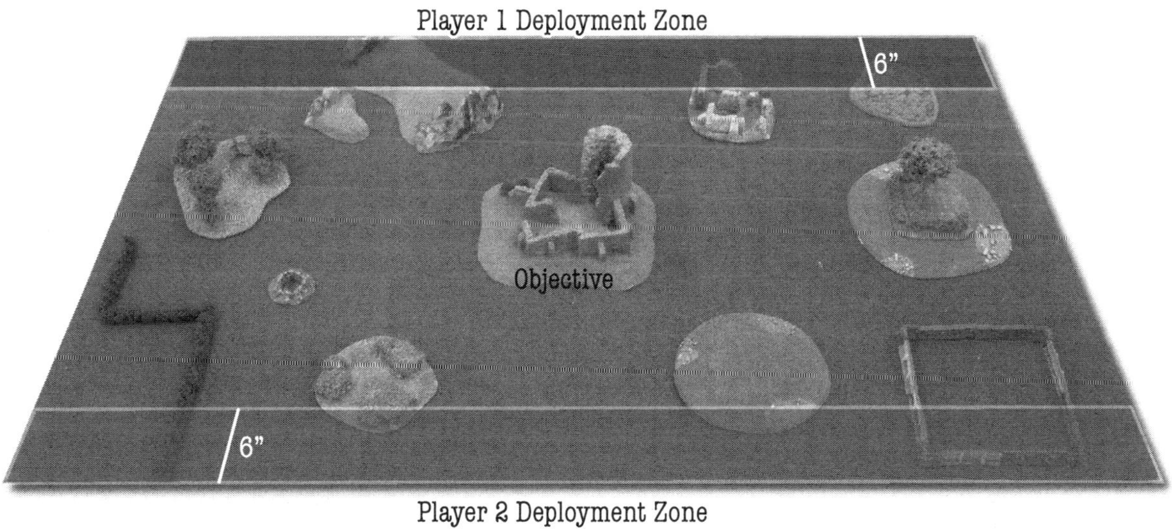

SWEEP AND CLEAR

The taking of territory is of prime importance in this battle, as the enemy must not only be defeated but beaten back so badly that he cannot counterattack. When two forces meet with the same objective, you can expect a grand battle of sweeping manoeuvres and vicious close-in fighting that employ the full capabilities of an army.

Briefing

High command has ordered you to take the ground on the far flank of the regiment. No enemy is to be left standing – enemy units are to be destroyed or routed with no quarter. Win this fight, and you will have achieved a tactical advantage in the battles to come.

- Roll a dice – if you roll an odd number, you must sweep the enemy out of the left half of the table. If you roll even, you must sweep the enemy out of the right half of the table.
- Roll a dice each – the lowest score will deploy first and take the first turn.

Scenario Rules

The scenario will last for six turns.

Victory Conditions

Victory Points are used in this scenario. Players lose Victory Points equal to the Value of each enemy unit still in the half of the table they were expected to sweep clear, no matter how many losses that unit has taken.

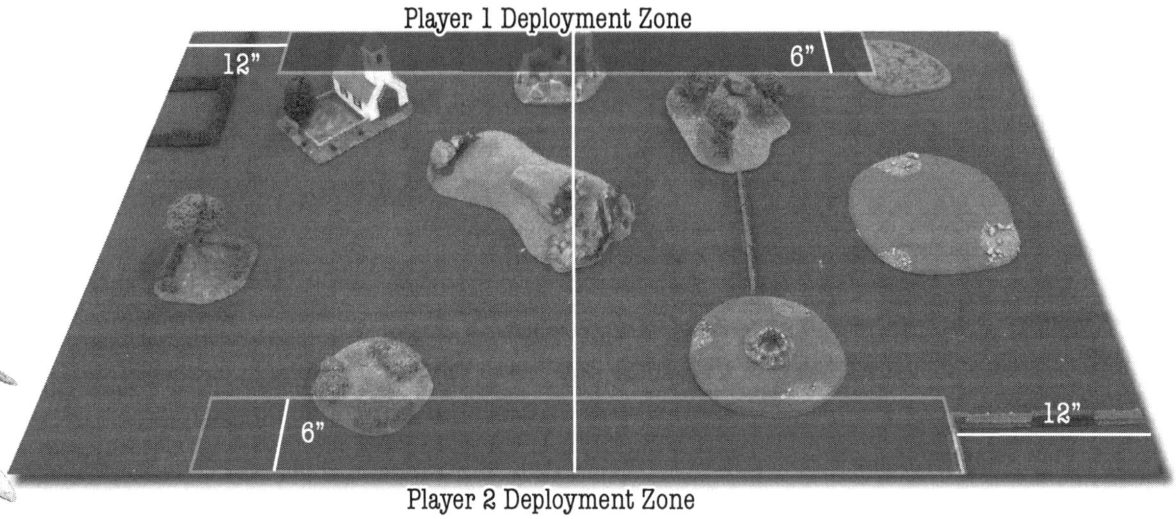

TOWERING INFERNO

Both sides have been searching for an opportunity to engage the enemy on a battlefield that will be to their advantage, engaging in small skirmishes as they seek to bring their main forces to bear. When the clash finally comes, no one is truly ready. Advanced units encounter the enemy and desperately call for reinforcements. Every available unit is diverted to the area in an attempt to gain control of the battlefield. Victory will lie with the side that can effectively stall the enemy while waiting for its own reserves to turn up.

Briefing

At last, the chance to get to grips with the enemy and deliver a final blow. Your forces are converging on this location already, all you need do is hold the enemy back long enough for your army to build up and overwhelm him.

- Roll a dice each – the lowest score will take the first turn.
- On your first turn, choose one of the units in your army. This will move on to the table from your table edge. Your opponent does likewise in his turn.
- The rest of your army enters as Reserves, but you have no control over when they appear and in what order. On your second turn, roll a D6 for every unit in your army. On a 6, the unit may be brought on to the table from your table edge.
- Repeat this every turn, but the number you need to roll for units to appear on the third turn is 5, on turn four it will be 4, and so on, until your entire army has moved on to the table.

Scenario Rules

The scenario will last for six turns.

Victory Conditions

Victory Points are used in this scenario.

Player 1 moves on
from this table edge

Player 2 moves on
from this table edge

URBAN RUMBLE

Battles in built up areas happened across both western and eastern fronts. Here, battles can become truly vicious, with forces fighting tooth and nail for each house, parade and square, edging forward into enemy held streets, inch by inch. Victory will belong not to the side who fields the heaviest weapons, but to the most disciplined and fearless troops.

Briefing

The enemy has deployed forces in this town, and it is up to you to force him out. You must steadily fight for each building, giving no quarter and without allowing the enemy to flank your troops. Armour and air power may prove beneficial in this environment, but they will not win the battle for you – instead, your best troops must be thrown into the meat grinder, in the hope that they will finish the enemy before the enemy shatters them.

- Roll a dice each – the lowest score will deploy first and take the first turn.

Scenario Rules

The scenario will last for twelve turns. The table is divided into 12" by 12" squares, each of which must have at least one structure placed within it. Each of these squares (and the streets and terrain within the square that surround the structure) counts as a single block. The two forces must fight for control of these blocks in order to win the battle.

Victory Conditions

Victory Points are used in this scenario. Both sides gain +250 bonus Victory Points for every block that they control at the end of the game. A block is controlled by having at least one friendly unit within the block (regardless of how many casualties it has taken), with no enemy units being present. Once a block has been taken in this fashion, it will remain in the player's control when his unit leaves, unless an enemy unit later takes control of it.

Player 2 Deployment zone

Player 1 Deployment zone

AIRBORNE ASSAULT

Paratroopers are used to spearhead a daring attack.. If the defender can repel the Paratroopers' attack the whole campaign of the attacker might fail.

Attacker's Briefing

Your army is on the move and to prepare the ground for your normal infantry forces your airborne platoon has to secure a vital position deep inside enemy territory. Your task is simple: take the objective and hold it until your ground forces arrive.

- Place a terrain piece or structure in anywhere on the table (a building or a bridge work well for this). This is the tactical objective you are fighting for.
- You may only include parachute deployable units or air units and may not include Emplacements.
- You do not deploy any units before the battle; you take the first turn and deploy your whole army by using the Parachute Deployment rule.
- You can choose to use the Night Fight rules for the scenario.

Defender's Briefing

A major enemy attack is about to come. Intelligence has warned you of a possible paratrooper attack by the enemy and have spread your forces thin to cover as much ground as possible. Hopefully your troops will repel the enemy's assault.

- Your army may not include Emplacements or any Air Units.
- Divide your forces in two halves with an equal as possible points value. You deploy only one half of you're your forces on the table before the battle. One of these units deployed before battle can be near the objective with the rest of the units deployed at least 16" away from the objective. You take the second turn.
- The rest of your army enters as Reserves, but you have no control over when they appear or in what order. On your second turn, roll a D6 for each Reserve unit. On a 6, the unit may be brought on to the table from a random table edge after rolling 1D6 as shown below. Repeat this every turn, but the number you need to roll for units to appear on the third turn is 5, on turn four it will be 4 and so on, until your entire army has moved on to the table from a random table edge.

Scenario Rules

The scenario will last for eight turns.

Victory Conditions

Victory Points are used in this scenario. Both sides gain bonus Victory Points equal to the Value of all of their units that are within 12" of the objective.

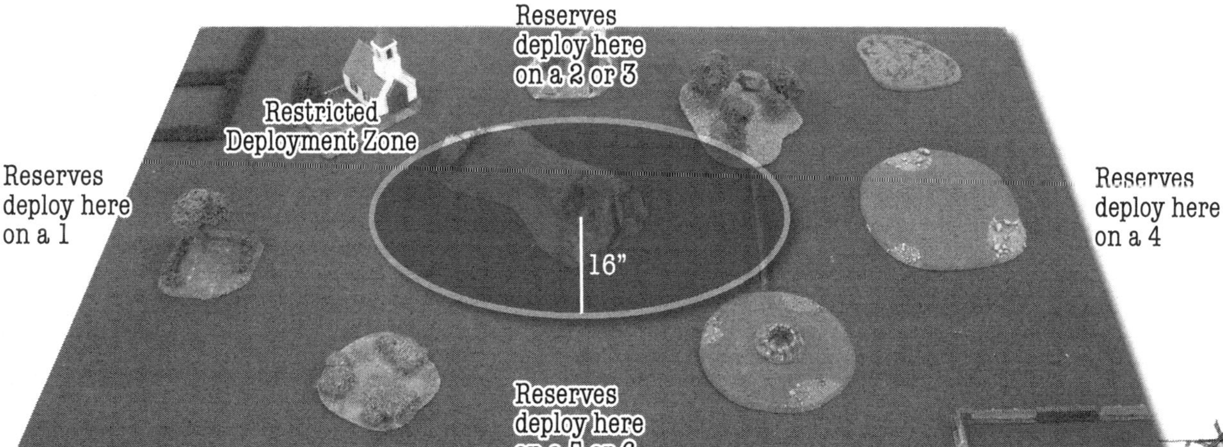